Student Resource and Solutions Manual

A First Course in Differential Equations with Modeling Applications

Ninth Edition

Dennis G. Zill
Loyola Marymount University

By

Warren S. Wright
Loyola Marymount University

Dennis G. Zill
Loyola Marymount University

Carol D. Wright

BROOKS/COLE
CENGAGE Learning

Australia • Brazil • Japan • Korea • Mexico • Singapore • Spain • United Kingdom • United States

BROOKS/COLE
CENGAGE Learning

For product information and technology assistance, contact us at
Cengage Learning Customer & Sales Support,
1-800-354-9706

For permission to use material from this text or product, submit all requests online at **www.cengage.com/permissions**
Further permissions questions can be emailed to
permissionrequest@cengage.com

ISBN-13: 978-0-495-38566-0
ISBN-10: 0-495-38566-2

Brooks/Cole
10 Davis Drive
Belmont, CA 94002-3098
USA

Cengage Learning is a leading provider of customized learning solutions with office locations around the globe, including Singapore, the United Kingdom, Australia, Mexico, Brazil, and Japan. Locate your local office at: **international.cengage.com/region**

Cengage Learning products are represented in Canada by Nelson Education, Ltd.

For your course and learning solutions, visit **academic.cengage.com**

Purchase any of our products at your local college store or at our preferred online store
www.ichapters.com

Printed in the United States of America
2 3 4 5 6 7 11

Introduction

This *Student Resource and Solutions Manual* is meant to accompany the text *A First Course in Differential Equations with Modeling Applications*, Ninth Edition. It is intended to serve several purposes:

- to provide a list of the terminology, concepts, and skills associated with each section in the text

- to provide further elaboration on some of the material discussed in the section

- to review some of the mathematics pertinent to the material being presented (such as integration, synthetic division, and partial fractions)

- to furnish extra examples of the material presented in the section

- to illustrate the use of the computer algebra systems *Mathematica* and *Maple* in dealing with the material in the section

- to provide solutions for every third problem in the exercise sets, excluding Discussion Problems, Mathematical Models, Computer Lab Assignments, Project Problems, and Contributed Problems

- to provide hints and suggestions for selected problems throughout the entire set of exercises in each section

- to provide examples illustrating how to solve some of the Computer Lab problems.

Table of Contents

1 Introduction to Differential Equations

Section 1.1
Definitions and Terminology

The terminology and concepts listed below provide an outline of the main ideas encountered in this section. These can be useful when preparing for a quiz or test.

Terminology and Concepts

- ordinary DE

- partial DE

- order of a DE

- normal form of a DE

- linear DE

- explicit and implicit form of a solution of a DE

- interval of definition of a solution

- trivial solution of a DE

- solution curve

- family of solutions of a DE

- n-parameter family of solutions of an nth-order DE

- particular solution of a DE

- singular solution of a DE

- system of DEs

- general solution of a DE

The basic skills listed below summarize the more mechanical types of problems encountered in the exercise set for this section.

Basic Skills

- determine the order of a DE
- determine if a DE is linear or nonlinear
- verify that a given function is a solution of a DE
- determine the interval of definition of a solution of a DE
- use an implicit solution to determine an explict solution (this is not always possible)
- identify constant solutions of certain DEs
- verify that a pair of functions is a solution of a system of two DEs

Basic Properties of the Derivative

Definition: The derivative of a function $y = f(x)$ with respect to the independent variable x is

$$f'(x) = \lim_{h \to 0} \frac{f(x+h) - f(x)}{h},$$

whenever this limit exists.

The derivative is also denoted by y' and dy/dx. When the derivative is evaluated at $x = x_0$, denoted by $f'(x_0)$, the resulting value represents the *rate of change* of the function f at x_0. Geometrically, this is the slope of the tangent line to the graph of $y = f(x)$ at the point $(x_0, f(x_0))$. If the function f is differentiable at x_0 then the graph of f at the point $(x_0, f(x_0))$ does not have a sharp corner or vertical tangent line at that point.

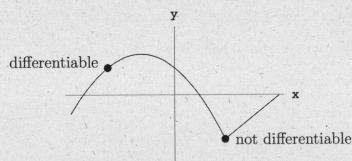

In DEs the independent variable frequently represents time and is denoted by t. In this case, if the position at time t of an object moving on a straight line is given by $s = s(t)$, then the velocity of the object at time t is the first derivative, $v(t) = s'(t)$, and the acceleration is the second derivative of $s(t)$, $a(t) = s''(t)$.

If the first derivative of a function is positive on an interval, then the function is **increasing** on that interval. Correspondingly, when the derivative is negative on an interval, the function is

decreasing on the interval. Relative extrema (maxima or minima) of a function occur where the derivative is either 0 or does not exist. The second derivative of a function can be used to determine concavity. The graph of a function f is **concave up** on an interval if $f''(x) > 0$ on the interval and **concave down** on an interval if $f''(x) < 0$ on the interval.

Hyperbolic Functions   It is unfortunate that many current texts and calculus instructors downplay the hyperbolic functions. These functions are important in applied mathematics and every serious student of differential equations should know their definitions, graphs, and some of their properties. In the text, solutions of DEs will be expressed in terms of hyperbolic functions when appropriate. The six hyperbolic functions are denoted by sinh, cosh, tanh, coth, sech and csch. If x denotes a real variable, the **hyperbolic sine** and **cosine** are defined in terms of exponential functions.

Definitions: $\sinh x = \dfrac{e^x - e^{-x}}{2}$ and $\cosh x = \dfrac{e^x + e^{-x}}{2}$ (1)

The remaining four hyperbolic functions are defined in a manner analogous to trigonometry. For example, the hyperbolic tangent and cotangent are, respectively,

$$\tanh x = \frac{\sinh x}{\cosh x} \quad \text{and} \quad \coth x = \frac{\cosh x}{\sinh x}.$$

The graphs of the hyperbolic sine and cosine functions, like the graphs of natural exponential, natural logarithmic, and trigonometric functions, should be memorized.

Graphs:

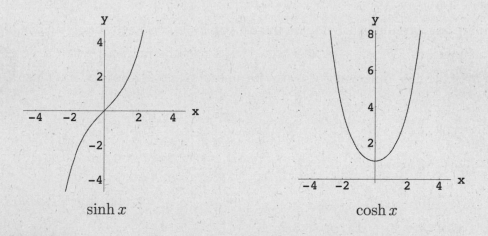

$\sinh x$ $\cosh x$

Some Properties   As can be seen from both the graphs and the definitions in (1),

$$\sinh 0 = 0 \quad \text{and} \quad \cosh 0 = 1.$$ (2)

Also, $\cosh x$ has no real zeros and, unlike the trigonometric functions, no hyperbolic function is periodic. However, a number of other properties similar to properties for trigonometric functions do hold. As you examine the hyperbolic identities (3) through (8) below, determine whether or not

3

the given identity differs from the analogous identity for trigonometric functions.

$$\cosh^2 x - \sinh^2 x = 1 \tag{3}$$

$$\sinh(x_1 \pm x_2) = \sinh x_1 \cosh x_2 \pm \cosh x_1 \sinh x_2 \tag{4}$$

$$\cosh(x_1 \pm x_2) = \cosh x_1 \cosh x_2 \pm \sinh x_1 \sinh x_2 \tag{5}$$

$$\sinh 2x = 2 \sinh x \cosh x \tag{6}$$

$$\cosh 2x = \cosh^2 x + \sinh^2 x \tag{7}$$

The derivatives of $\sinh x$ and $\cosh x$, given below, can be obtained using the definitions in (1).

$$\frac{d}{dx} \sinh x = \cosh x \qquad \text{and} \qquad \frac{d}{dx} \cosh x = \sinh x \tag{8}$$

Use of Computers To define a function, such as $f(x) = e^{x^2} + \sin 2x$, use the syntax

```
Clear[f]                                    (Mathematica)
f[x_]:= Exp[x^2]+Sin[2x]
f[x]

f:=x->exp(x^2)+sin(2*x);                     (Maple)
or
f:=unapply(exp(x^2)+sin(2*x), x);
```

If you are not familiar with *Maple* the **unapply** function probably seems like an awkward way to define a function. We will see later that it is a useful command when solving DEs.

Piecewise-defined functions are, as the name suggests, defined in pieces. For example, the function

$$f(x) = \begin{cases} x^2 + 1, & x \le 2 \\ \sin x, & x > 2 \end{cases}$$

is defined in a CAS using:

```
Clear[f]                                    (Mathematica)
f[x_]:= x^2 + 1/; x <= 2
f[x_]:= Sin[x]/; x > 2

f:=x->piecewise(x<=2, x^2+1, x>2,sin(x));   (Maple)
```

The syntax for the first derivative, dy/dx, of $y = f(x)$ is:

```
D[f[x], x]  or  f'[x]                        (Mathematica)
```

```
diff(f(x), x);
```
(Maple)

The second derivative, d^2y/dx^2, of $y = f(x)$ is:

$$D[f[x], \{x,2\}] \quad \text{or} \quad f''[x]$$
(Mathematica)

```
diff(f(x), x$2);   or   diff(f(x),x,x);
```
(Maple)

Exercises 1.1   *Hints, Suggestions, Solutions, and Examples*

3. Fourth order; linear

6. Second order; nonlinear because of R^2

9. Writing the DE in the form $x(dy/dx) + y^2 = 1$, we see that it is nonlinear in y because of y^2. However, writing it in the form $(y^2 - 1)(dx/dy) + x = 0$, we see that it is linear in x.

12. From $y = \frac{6}{5} - \frac{6}{5}e^{-20t}$ we obtain $dy/dt = 24e^{-20t}$, so that

$$\frac{dy}{dt} + 20y = 24e^{-20t} + 20\left(\frac{6}{5} - \frac{6}{5}e^{-20t}\right) = 24.$$

15. The domain of the function, found by solving $x + 2 \geq 0$, is $[-2, \infty)$. From $y' = 1 + 2(x+2)^{-1/2}$ we have

$$(y - x)y' = (y - x)[1 + (2(x+2)^{-1/2}]$$

$$= y - x + 2(y - x)(x+2)^{-1/2}$$

$$= y - x + 2[x + 4(x+2)^{1/2} - x](x+2)^{-1/2}$$

$$= y - x + 8(x+2)^{1/2}(x+2)^{-1/2} = y - x + 8.$$

An interval of definition for the solution of the DE is $(-2, \infty)$ because y' is not defined at $x = -2$.

18. The function is $y = 1/\sqrt{1 - \sin x}$, whose domain is obtained from $1 - \sin x \neq 0$ or $\sin x \neq 1$. Thus, the domain is $\{x \mid x \neq \pi/2 + 2n\pi\}$. From $y' = -\frac{1}{2}(1 - \sin x)^{-3/2}(-\cos x)$ we have

$$2y' = (1 - \sin x)^{-3/2}\cos x = [(1 - \sin x)^{-1/2}]^3 \cos x = y^3 \cos x.$$

An interval of definition for the solution of the DE is $(\pi/2, 5\pi/2)$. Another interval is $(5\pi/2, 9\pi/2)$ and so on.

21. Differentiating $P = c_1 e^t / \left(1 + c_1 e^t\right)$ we obtain

$$\frac{dP}{dt} = \frac{\left(1 + c_1 e^t\right)c_1 e^t - c_1 e^t \cdot c_1 e^t}{(1 + c_1 e^t)^2} = \frac{c_1 e^t}{1 + c_1 e^t}\frac{\left[\left(1 + c_1 e^t\right) - c_1 e^t\right]}{1 + c_1 e^t}$$

$$= \frac{c_1 e^t}{1 + c_1 e^t}\left[1 - \frac{c_1 e^t}{1 + c_1 e^t}\right] = P(1 - P).$$

22. (*Hint*) Assume a function $y = f(x)$ is continuous on $[a, b]$, and that x is any number in the interval. The derivative form of the **Fundamental Theorem of Calculus** is then

$$\frac{d}{dx} \int_a^x f(t)\, dt = f(x).$$

Using the Chain Rule we then have

$$\frac{d}{dx} \int_a^{g(x)} f(t)\, dt = f(g(x))g'(x).$$

24. From $y = c_1 x^{-1} + c_2 x + c_3 x \ln x + 4x^2$ we obtain

$$\frac{dy}{dx} = -c_1 x^{-2} + c_2 + c_3 + c_3 \ln x + 8x,$$

$$\frac{d^2 y}{dx^2} = 2c_1 x^{-3} + c_3 x^{-1} + 8,$$

and

$$\frac{d^3 y}{dx^3} = -6c_1 x^{-4} - c_3 x^{-2},$$

so that

$$x^3 \frac{d^3 y}{dx^3} + 2x^2 \frac{d^2 y}{dx^2} - x \frac{dy}{dx} + y = (-6c_1 + 4c_1 + c_1 + c_1)x^{-1} + (-c_3 + 2c_3 - c_2 - c_3 + c_2)x$$

$$+ (-c_3 + c_3)x \ln x + (16 - 8 + 4)x^2$$

$$= 12x^2.$$

25. (*Hint*) For a piecewise-defined function to be smooth at the point where the pieces are joined the function must be continuous there, and the left-hand and right-hand derivatives of the two pieces at the point must be equal.

27. From $y = e^{mx}$ we obtain $y' = me^{mx}$. Then $y' + 2y = 0$ implies

$$me^{mx} + 2e^{mx} = (m + 2)e^{mx} = 0.$$

Since $e^{mx} > 0$ for all x, $m = -2$. Thus $y = e^{-2x}$ is a solution.

30. From $y = e^{mx}$ we obtain $y' = me^{mx}$ and $y'' = m^2 e^{mx}$. Then $2y'' + 7y' - 4y = 0$ implies

$$2m^2 e^{mx} + 7me^{mx} - 4e^{mx} = (2m - 1)(m + 4)e^{mx} = 0.$$

Since $e^{mx} > 0$ for all x, $m = \frac{1}{2}$ and $m = -4$. Thus $y = e^{x/2}$ and $y = e^{-4x}$ are solutions.

In Problems 33–36 we substitute $y = c$ into the DEs and use $y' = 0$ and $y'' = 0$.

33. Solving $5c = 10$ we see that $y = 2$ is a constant solution.

36. Solving $6c = 10$ we see that $y = 5/3$ is a constant solution.

45-46. (*Hint*) A function is not differentiable at a point where a tangent line to its graph is vertical.

54. (*Suggestion*) You have y, y', and y''. Eliminate c_1 and c_2 between these expressions.

55-58. (*Suggestion*) In this manual, see **Signs of Derivatives** at the beginning of Section 2.1.

59-60. (*Example*) Use a CAS to verify that $y = 4x^2 \sin(5 \ln x) + 13x \ln x + x$ is a solution of the DE $x^2 y'' - 3xy' + 29y = 338x \ln x$.

```
Clear[y]                                        (Mathematica)
y[x_]:= 4x^2 Sin[5Log[x]] + 13x Log[x] + x
y[x]

x^2 y''[x] - 3x y'[x] + 29y[x] // Simplify
```

```
y := x->4*x^2*sin(5*ln(x))+13*x*ln(x)+x;        (Maple)
x^2*diff(y(x),x$2)-3*x*diff(y(x),x)+29*y(x);
simplify(%);
```

In both cases the output of the simplified expression is $338x \ln x$, which verifies that the given function is a solution of the DE for $x > 0$.

Section 1.2
Initial-Value Problems

The terminology and concepts listed below provide an outline of the main ideas encountered in this section. These can be useful when preparing for a quiz or test.

Terminology and Concepts

- IVPs

- initial conditions

- how the interval of definition of a solution of an IVP may differ from the domain of the function in the solution

- existence and uniqueness of solutions of IVPs

The basic skills listed below summarize the more mechanical types of problems encountered in the exercise set for this section.

Basic Skills

- given a family of solutions of a DE, find a particular solution satisfying given initial conditions

- determine a region in the plane for which a given DE will have a unique solution whose graph passes through a point in the region

- match graphs of solution curves with given initial conditions

Existence and Uniqueness
Theorem 1.2.1 is local in nature, so the interval I_0 may not be very large. The words "existence" and "uniqueness" are often not clearly understood. Use of the notation such as $y = y(x)$ and saying that a solution "exists" on the interval I_0 does *not* necessarily mean that we can actually find (meaning display) an explicit or implicit solution of the DE defined on I_0. We know in theory that a solution $y = y(x)$ exists but we may have to approximate it on I_0. The word "uniqueness" basically means we cannot have *two* (or more) solutions $y = y_1(x)$ and $y = y_2(x)$ defined on I_0 such that their solution curves both pass through the initial point (x_0, y_0). In other words, on I_0 we cannot have either of the following situations:

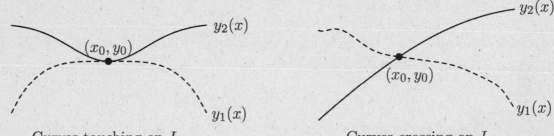

Curves touching on I_0 Curves crossing on I_0

Since any point on the graph of a solution curve may be considered as an initial condition of the DE resulting in that curve, no two solution curves can intersect.

Exercises 1.2   *Hints, Suggestions, Solutions, and Examples*

3. Letting $x = 2$ and solving $1/3 = 1/(4 + c)$ we get $c = -1$. The solution is $y = 1/(x^2 - 1)$. This solution is defined on the interval $(1, \infty)$.

6. Letting $x = 1/2$ and solving $-4 = 1/(1/4 + c)$ we get $c = -1/2$. The solution is $y = 1/(x^2 - 1/2) = 2/(2x^2 - 1)$. This solution is defined on the interval $(-1/\sqrt{2}, 1/\sqrt{2})$.

9. In this problem we use $x = c_1 \cos t + c_2 \sin t$ and $x' = -c_1 \sin t + c_2 \cos t$ to obtain a system of two equations in the two unknowns c_1 and c_2. From the initial conditions we obtain

$$\frac{\sqrt{3}}{2} c_1 + \frac{1}{2} c_2 = \frac{1}{2}$$

$$-\frac{1}{2} c_1 + \frac{\sqrt{3}}{2} c_2 = 0.$$

Solving, we find $c_1 = \sqrt{3}/4$ and $c_2 = 1/4$. The solution of the IVP is $x = (\sqrt{3}/4) \cos t + (1/4) \sin t$.

11-14. (*Suggestion*) Another two-parameter family of solutions of the DE in these problems is given by $y = c_3 \cosh x + c_4 \sinh x$. You are encouraged to find solutions of these IVPs and then show that these solutions are equivalent to the ones obtained from the family $y = c_1 e^x + c_2 e^{-x}$.

12. In this problem we use $y = c_1 e^x + c_2 e^{-x}$ and $y' = c_1 e^x - c_2 e^{-x}$ to obtain a system of two equations in the two unknowns c_1 and c_2. From the initial conditions we obtain

$$ec_1 + e^{-1} c_2 = 0$$

$$ec_1 - e^{-1} c_2 = e.$$

Solving, we find $c_1 = \frac{1}{2}$ and $c_2 = -\frac{1}{2} e^2$. The solution of the IVP is $y = \frac{1}{2} e^x - \frac{1}{2} e^2 e^{-x} = \frac{1}{2} e^x - \frac{1}{2} e^{2-x}$.

15. Two solutions are $y = 0$ and $y = x^3$.

18. For $f(x, y) = \sqrt{xy}$ we have $\partial f/\partial y = \frac{1}{2} \sqrt{x/y}$. Thus, the DE will have a unique solution in any region where $x > 0$ and $y > 0$ or where $x < 0$ and $y < 0$.

21. For $f(x, y) = x^2/(4 - y^2)$ we have $\partial f/\partial y = 2x^2 y/(4 - y^2)^2$. Thus the DE will have a unique solution in any region where $y < -2$, $-2 < y < 2$, or $y > 2$.

24. For $f(x, y) = (y + x)/(y - x)$ we have $\partial f/\partial y = -2x/(y - x)^2$. Thus the DE will have a unique solution in any region where $y < x$ or where $y > x$.

27. Identify $f(x, y) = \sqrt{y^2 - 9}$ and $\partial f/\partial y = y/\sqrt{y^2 - 9}$. We see that f and $\partial f/\partial y$ are both continuous in the regions of the plane determined by $y < -3$ and $y > 3$ with no restrictions on x. Since $(2, -3)$ is not in either of these regions, there is no guarantee of a unique solution through $(2, -3)$.

30. (a) Since $\dfrac{d}{dx} \tan(x + c) = \sec^2(x + c) = 1 + \tan^2(x + c)$, we see that $y = \tan(x + c)$ satisfies the DE.

(b) Solving $y(0) = \tan c = 0$ we obtain $c = 0$ and $y = \tan x$. Since $\tan x$ is discontinuous at $x = \pm\pi/2$, the solution is not defined on $(-2, 2)$ because it contains $\pm\pi/2$.

(c) The largest interval on which the solution can exist is $(-\pi/2, \pi/2)$.

33. (a) Differentiating $3x^2 - y^2 = c$ we get $6x - 2yy' = 0$ or $yy' = 3x$.

(b) Solving $3x^2 - y^2 = 3$ for y we get

$$y = \phi_1(x) = \sqrt{3(x^2 - 1)}, \qquad 1 < x < \infty,$$

$$y = \phi_2(x) = -\sqrt{3(x^2 - 1)}, \qquad 1 < x < \infty,$$

$$y = \phi_3(x) = \sqrt{3(x^2 - 1)}, \qquad -\infty < x < -1,$$

$$y = \phi_4(x) = -\sqrt{3(x^2 - 1)}, \qquad -\infty < x < -1.$$

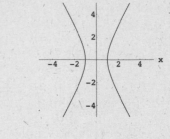

(c) Only $y = \phi_3(x)$ satisfies $y(-2) = 3$.

35-41. (*Suggestion*) For problems 35–41 you might find it helpful to read the comments **Signs of Derivatives** at the beginning of Section 2.1.

36. From the graph we see that $y(0) = -1$, so the only possible initial conditions are those in (d) and (e). Again from the graph, at $x = 0$, the tangent line is horizontal, so $y'(0) = 0$, and the graph corresponds to the initial conditions in (e).

Section 1.3

Differential Equations as Mathematical Models

The terminology and concepts listed below provide an outline of the main ideas encountered in this section. These can be useful when preparing for a quiz or test.

Terminology and Concepts

- construction of mathematical models
- level of resolution of a model
- use of assumptions and empirical laws
- steps of the modeling process
- model for population growth
- model for radioactive decay
- Newton's law of cooling/warming
- model for the spread of a disease
- model for the draining of a tank

- models for first- and second-order chemical reactions
- models to describe mixtures of differing concentrations of elements in liquid solutions
- models for electrical series circuits
- models for the motion of falling bodies involving Newton's laws of motion
- approaches to DEs: analytical, qualitative, numerical

The basic skills listed below summarize the more mechanical types of problems encountered in the exercise set for this section.

Basic Skills

- set up DEs, sometimes with initial conditions, that can be used to model the situations listed above

Mixtures   In setting up DEs such as

$$\frac{dA}{dt} + \frac{1}{100} A = 6 \qquad \text{[(8) in the text]}$$

it is helpful to keep in mind the units. The derivative, dA/dt, is the rate at which the amount of salt (pounds) is changing in time (minutes) and so is measured in pounds per minute. In the derivation of (8) in the text observe that the unit "gal" cancels out in both R_{in} and R_{out} to give "lb/min." Also, note carefully that even though the tank contains 300 gallons of brine, the initial condition associated with (8) is *not* $A(0) = 300$ since $A(t)$, which represents the amount of salt at time t, is measured in pounds, whereas 300 is a measure of the number of gallons of fluid in the tank. Indeed, an initial condition for $A(t)$ is not specified. See the text: Problem 9 in Exercises 1.3.

Exercises 1.3   *Hints, Suggestions, Solutions, and Examples*

3. Let b be the rate of births and d the rate of deaths. Then $b = k_1 P$ and $d = k_2 P^2$. Since $dP/dt = b - d$, the DE is $dP/dt = k_1 P - k_2 P^2$.

6. By inspecting the graph in the text we take T_m to be $T_m(t) = 80 - 30 \cos \pi t/12$. Then the temperature of the body at time t is determined by the DE

$$\frac{dT}{dt} = k \left[T - \left(80 - 30 \cos \frac{\pi}{12} t \right) \right], \quad t > 0.$$

8. (*Hint*) A quantity P is jointly proportional to two quantities x and y if there exists a constant k such that $P = kxy$.

9. The rate at which salt is leaving the tank is

$$R_{out} \ (3 \text{ gal/min}) \cdot \left(\frac{A}{300} \text{ lb/gal}\right) = \frac{A}{100} \text{ lb/min}.$$

Thus $dA/dt = -A/100$ (where the minus sign is used since the amount of salt is decreasing. The initial amount is $A(0) = 50$.

12. The rate at which salt is entering the tank is

$$R_{in} = (c_{in} \text{ lb/gal}) \cdot (r_{in} \text{ gal/min}) = c_{in}r_{in} \text{ lb/min}.$$

Now let $A(t)$ denote the number of pounds of salt and $N(t)$ the number of gallons of brine in the tank at time t. The concentration of salt in the tank as well as in the outflow is $c(t) = x(t)/N(t)$. But the number of gallons of brine in the tank remains steady, is increased, or is decreased depending on whether $r_{in} = r_{out}$, $r_{in} > r_{out}$, or $r_{in} < r_{out}$. In any case, the number of gallons of brine in the tank at time t is $N(t) = N_0 + (r_{in} - r_{out})t$. The output rate of salt is then

$$R_{out} = \left(\frac{A}{N_0 + (r_{in} - r_{out})t} \text{ lb/gal}\right) \cdot (r_{out} \text{ gal/min}) = r_{out} \frac{A}{N_0 + (r_{in} - r_{out})t} \text{ lb/min}.$$

The DE for the amount of salt, $dA/dt = R_{in} - R_{out}$, is

$$\frac{dA}{dt} = c_{in}r_{in} - r_{out}\frac{A}{N_0 + (r_{in} - r_{out})t} \quad \text{or} \quad \frac{dA}{dt} + \frac{r_{out}}{N_0 + (r_{in} - r_{out})t}A = c_{in}r_{in}.$$

15. Since $i = dq/dt$ and $L\,d^2q/dt^2 + R\,dq/dt = E(t)$, we obtain $L\,di/dt + Ri = E(t)$.

18. Since the barrel in Figure 1.3.16(b) in the text is submerged an additional y feet below its equilibrium position the number of cubic feet in the additional submerged portion is the volume of the circular cylinder: $\pi \times (\text{radius})^2 \times \text{height}$ or $\pi(s/2)^2y$. Then we have from Archimedes' principle

$$\text{upward force of water on barrel} = \text{weight of water displaced}$$

$$= (62.4) \times (\text{volume of water displaced})$$

$$= (62.4)\pi(s/2)^2y = 15.6\pi s^2y.$$

It then follows from Newton's second law that

$$\frac{w}{g}\frac{d^2y}{dt^2} = -15.6\pi s^2y \quad \text{or} \quad \frac{d^2y}{dt^2} + \frac{15.6\pi s^2 g}{w}y = 0,$$

where $g = 32$ and w is the weight of the barrel in pounds.

20. (*Hint*) If the damping force is proportional to the instantaneous velocity, then there exists a constant β such that this force is equal to $\beta\,dx/dt$.

21. From $g = k/R^2$ we find $k = gR^2$. Using $a = d^2r/dt^2$ and the fact that the positive direction is upward we get

$$\frac{d^2r}{dt^2} = -a = -\frac{k}{r^2} = -\frac{gR^2}{r^2} \quad \text{or} \quad \frac{d^2r}{dt^2} + \frac{gR^2}{r^2} = 0.$$

24. The DE is $\dfrac{dA}{dt} = k_1(M - A) - k_2A$.

27. We see from the figure that $2\theta + \alpha = \pi$. Thus

$$\frac{y}{-x} = \tan\alpha = \tan(\pi - 2\theta) = -\tan 2\theta = -\frac{2\tan\theta}{1 - \tan^2\theta}.$$

Since the slope of the tangent line is $y' = \tan\theta$ we have $y/x = 2y'/[1 - (y')^2]$ or $y - y(y')^2 = 2xy'$, which is the quadratic equation $y(y')^2 + 2xy' - y = 0$ in y'. Using the quadratic formula, we get

$$y' = \frac{-2x \pm \sqrt{4x^2 + 4y^2}}{2y} = \frac{-x \pm \sqrt{x^2 + y^2}}{y}.$$

Since $dy/dx > 0$, the DE is

$$\frac{dy}{dx} = \frac{-x + \sqrt{x^2 + y^2}}{y} \qquad \text{or} \qquad y\frac{dy}{dx} - \sqrt{x^2 + y^2} + x = 0.$$

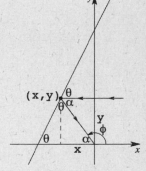

Chapter 1 in Review

Hints, Suggestions, Solutions, and Examples

3. $\dfrac{d}{dx}(c_1\cos kx + c_2\sin kx) = -kc_1\sin kx + kc_2\cos kx;$

$\dfrac{d^2}{dx^2}(c_1\cos kx + c_2\sin kx) = -k^2c_1\cos kx - k^2c_2\sin kx = -k^2(c_1\cos kx + c_2\sin kx);$

$\dfrac{d^2y}{dx^2} = -k^2y \quad \text{or} \quad \dfrac{d^2y}{dx^2} + k^2y = 0$

6. $y' = -c_1e^x\sin x + c_1e^x\cos x + c_2e^x\cos x + c_2e^x\sin x;$

$y'' = -c_1e^x\cos x - c_1e^x\sin x - c_1e^x\sin x + c_1e^x\cos x - c_2e^x\sin x + c_2e^x\cos x + c_2e^x\cos x + c_2e^x\sin x$

$\quad = -2c_1e^x\sin x + 2c_2e^x\cos x;$

$y'' - 2y' = -2c_1e^x\cos x - 2c_2e^x\sin x = -2y; \qquad y'' - 2y' + 2y = 0$

In Problems 9 and 12 the possible solutions can simply be tested in the given DEs.

9. b

12. a,b,d

15. The slope of the tangent line at (x, y) is y', so the DE is $y' = x^2 + y^2$.

18. **(a)** Differentiating $y^2 - 2y = x^2 - x + c$ we obtain $2yy' - 2y' = 2x - 1$ or $(2y - 2)y' = 2x - 1$.

(b) Setting $x = 0$ and $y = 1$ in the solution we have $1 - 2 = 0 - 0 + c$ or $c = -1$. Thus, a solution of the IVP is $y^2 - 2y = x^2 - x - 1$.

(c) Solving $y^2 - 2y - (x^2 - x - 1) = 0$ by the quadratic formula we get $y = (2 \pm \sqrt{4 + 4(x^2 - x - 1)})/2$
$= 1 \pm \sqrt{x^2 - x} = 1 \pm \sqrt{x(x - 1)}$. Since $x(x - 1) \geq 0$ for $x \leq 0$ or $x \geq 1$, we see that neither $y = 1 + \sqrt{x(x - 1)}$ nor $y = 1 - \sqrt{x(x - 1)}$ is differentiable at $x = 0$. Thus, both functions are solutions of the DE, but neither is a solution of the IVP.

21. (a)

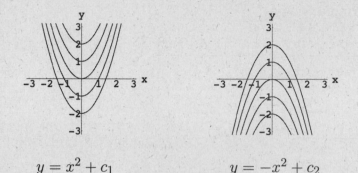

$$y = x^2 + c_1 \qquad\qquad y = -x^2 + c_2$$

(b) When $y = x^2 + c_1$, $y' = 2x$ and $(y')^2 = 4x^2$. When $y = -x^2 + c_2$, $y' = -2x$ and $(y')^2 = 4x^2$.

(c) Pasting together x^2, $x \geq 0$, and $-x^2$, $x \leq 0$, we get $y = \begin{cases} -x^2, & x \leq 0 \\ x^2, & x > 0. \end{cases}$

24. Differentiating $y = x \sin x + (\cos x) \ln(\cos x)$ we get

$$y' = x \cos x + \sin x + \cos x \left(\frac{-\sin x}{\cos x} \right) - (\sin x) \ln(\cos x)$$

$$= x \cos x + \sin x - \sin x - (\sin x) \ln(\cos x)$$

$$= x \cos x - (\sin x) \ln(\cos x)$$

and

$$y'' = -x \sin x + \cos x - \sin x \left(\frac{-\sin x}{\cos x} \right) - (\cos x) \ln(\cos x)$$

$$= -x \sin x + \cos x + \frac{\sin^2 x}{\cos x} - (\cos x) \ln(\cos x)$$

$$= -x \sin x + \cos x + \frac{1 - \cos^2 x}{\cos x} - (\cos x) \ln(\cos x)$$

$$= -x \sin x + \cos x + \sec x - \cos x - (\cos x) \ln(\cos x)$$

$$= -x \sin x + \sec x - (\cos x) \ln(\cos x).$$

Thus

$$y'' + y = -x \sin x + \sec x - (\cos x) \ln(\cos x) + x \sin x + (\cos x) \ln(\cos x) = \sec x.$$

To obtain an interval of definition we note that the domain of $\ln x$ is $(0, \infty)$, so we must have $\cos x > 0$. Thus, an interval of definition is $(-\pi/2, \pi/2)$.

In Problems 27 and 30 we have $y' = 3c_1 e^{3x} - c_2 e^{-x} - 2$.

27. The initial conditions imply

$$c_1 + c_2 = 0$$

$$3c_1 - c_2 - 2 = 0,$$

so $c_1 = \frac{1}{2}$ and $c_2 = -\frac{1}{2}$. Thus $y = \frac{1}{2}e^{3x} - \frac{1}{2}e^{-x} - 2x$.

30. The initial conditions imply

$$c_1 e^{-3} + c_2 e + 2 = 0$$

$$3c_1 e^{-3} - c_2 e - 2 = 1,$$

so $c_1 = \frac{1}{4}e^3$ and $c_2 = -\frac{9}{4}e^{-1}$. Thus $y = \frac{1}{4}e^{3x+3} - \frac{9}{4}e^{-x-1} - 2x$.

33. Let $P(t)$ be the number of owls present at time t. Then $dP/dt = k(P - 200 + 10t)$.

2 First-Order Differential Equations

Solution Curves Without a Solution

The terminology and concepts listed below provide an outline of the main ideas encountered in this section. These can be useful when preparing for a quiz or test.

Terminology and Concepts

- lineal element
- direction field (slope field)
- autonomous first-order DE
- critical (stationary, equilibrium) point
- equilibrium solution
- phase portrait (phase line)
- asymptotically stable critical point (attractor)
- unstable critical point (repeller)
- semi-stable critical point

The basic skills listed below summarize the more mechanical types of problems encountered in the exercise set for this section.

Basic Skills

- sketch a direction field of a first-order DE by hand
- sketch a solution curve through a given point in a direction field
- find the critical points of an autonomous first-order DE
- draw the phase portrait of an autonomous first-order DE
- classify critical points as asymptotically stable (attractor), unstable (repeller) or semi-stable
- use a phase portrait to sketch solution curves of autonomous first-order DEs

- determine the behavior of solutions of autonomous first-order DEs as $t \to \pm\infty$

Signs of Derivatives

The first derivative and its algebraic sign over an interval plays an important role in the qualitative analysis in Section 2.1. If $y = f(x)$ is a differentiable function and if $dy/dx > 0$ on an interval then f is increasing on the interval, whereas if $dy/dx < 0$ on an interval then f is decreasing on the interval. Similarly, if $y = f(x)$ possesses a second derivative, then the sign of d^2y/dx^2 determines the concavity of the graph of f. If $d^2y/dx^2 > 0$ on an interval then the graph of f is concave upward on the interval, whereas if $d^2y/dx^2 < 0$ on an interval, then the graph of f is concave downward.

Use of Computers

Both *Mathematica* and *Maple* can be used to obtain direction fields of first-order DEs. In *Mathematica*, a graphics package containing the command **PlotVectorField** must first be loaded. (It is important that this be done only once during any session of *Mathematica* and that it be done before the command is first used. If this protocol is not followed, you should save your current file, quit *Mathematica*, and restart the program.) **PlotVectorField** is actually used to graph the vector field corresponding to a vector function so it must be adapted to the case of a first-order DE having the form $dy/dx = f(x, y)$. The routine below plots the direction field for the DE $dy/dx = 0.2xy$ where $-5 < x < 5$ and $-5 < y < 5$.

```
<<Graphics`                                          (Mathematica)
PlotVectorField[{1, 2x y}, {x, -5, 5}, {y, -5, 5},
     Frame->True, Axes->True, AxesLabel->{"x", "y"},
     Plotpoints->15]
```

This example will plot a direction field in which the lineal elements (arrows) have their base at a grid point. The number of grid points in both horizontal and vertical directions is specified by the **PlotPoints** option. The output is shown below in Figure 2.1. To obtain a direction field with linear elements all the same length, the graphics options **ScaleFunction** and **ScaleFactor** can be added as shown below.

```
PlotVectorField[{1, 2x y},                           (Mathematica)
     {x, -5, 5}, {y, -5, 5}, Frame->True, Axes->True,
     AxesLabel->{"x", "y"}, ScaleFunction->(0.5&),
     ScaleFactor->0.5, PlotPoints->9]
```

The output for this command is shown below in Figure 2.2.

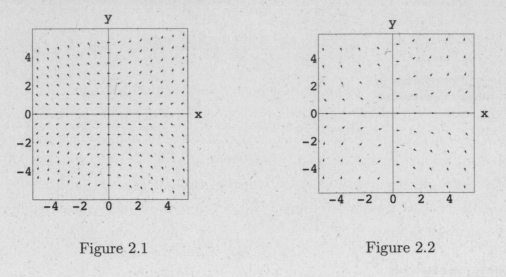

Figure 2.1   Figure 2.2

In *Maple* the command `fieldplot` is used to obtain a direction field. The routine below plots the direction field for the DE $dy/dx = 0.2xy$.

```
with(plots);                                    (Maple)
fieldplot([1, 0.2*x*y], x=-5..5, y=-5..5);
```

To obtain a direction field with lineal elements all having the same length it is necessary to scale the input by the length of each vector as shown below.

```
with(plots);                                    (Maple)
fieldplot([1/sqrt(1+(0.2*x*y)^2), 0.2*x*y/sqrt(1+(0.2*x*y)^2)],
    x=-5..5, y=-5..5);
```

Exercises 2.1   *Hints, Suggestions, Solutions, and Examples*

3.   6.

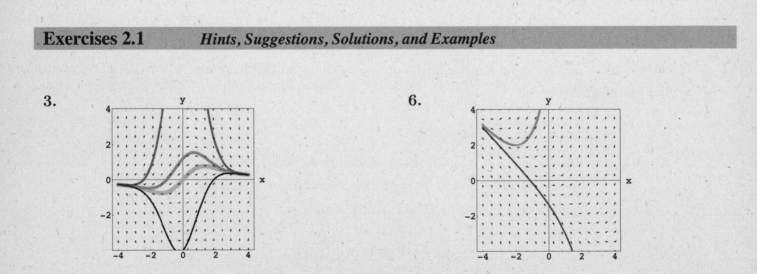

9.

12.

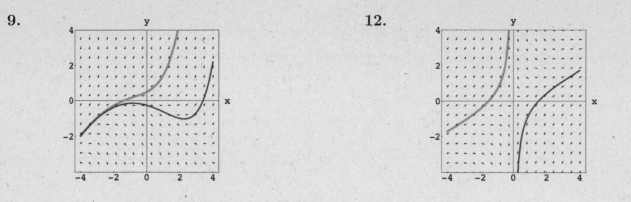

13. (*Hint*) The slope of a lineal element at the point (x_1, y_1) in a grid is $f(y_1)$ and can be approximated by measuring up from the y-axis in Figure 2.1.15 in the text.

15. (a) The isoclines have the form $y = -x + c$, which are straight lines with slope -1.

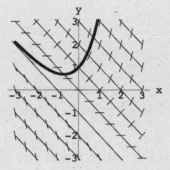

(b) The isoclines have the form $x^2 + y^2 = c$, which are circles centered at the origin.

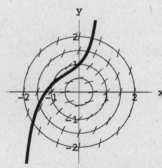

21. Solving $y^2 - 3y = y(y-3) = 0$ we obtain the critical points 0 and 3. From the phase portrait we see that 0 is asymptotically stable (attractor) and 3 is unstable (repeller).

24. Solving $10 + 3y - y^2 = (5 - y)(2 + y) = 0$ we obtain the critical points -2 and 5. From the phase portrait we see that 5 is asymptotically stable (attractor) and -2 is unstable (repeller).

27. Solving $y \ln(y + 2) = 0$ we obtain the critical points -1 and 0. From the phase portrait we see that -1 is asymptotically stable (attractor) and 0 is unstable (repeller).

30. The critical points are approximately at $-2, 2,\ 0.5,$ and 1.7. Since $f(y) > 0$ for $y < -2.2$ and $0.5 < y < 1.7$, the graph of the solution is increasing on $(-\infty, -2.2)$ and $(0.5, 1.7)$. Since $f(y) < 0$ for $-2.2 < y < 0.5$ and $y > 1.7$, the graph is decreasing on $(-2.2, 0.5)$ and $(1.7, \infty)$.

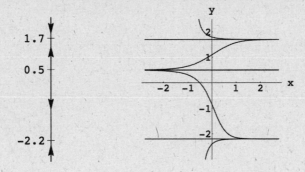

34. (*Suggestion*) Start by assuming that $\lim\limits_{x \to \infty} \dfrac{dy}{dx} = f(\lim\limits_{x \to \infty} y)$.

Section 2.2 Separable Variables

The terminology and concepts listed below provide an outline of the main ideas encountered in this section. These can be useful when preparing for a quiz or test.

Terminology and Concepts

- separable DE
- solutions with nonelementary functions

The basic skills listed below summarize the more mechanical types of problems encountered in the exercise set for this section.

Basic Skills

- determine whether or not a DE is separable
- use integration to solve a separable DE
- when possible, solve an equation for the dependent variable to find an explicit solution of a DE

Integration   The method of separation of variables is the first of several techniques for solving DEs that require integration. Recall from calculus that an indefinite integral (or antiderivative) of a function $f(x)$ is another function $F(x)$ such that $F'(x) = f(x)$. The most general indefinite integral of f is written:

$$\int f(x)dx = F(x) + c, \tag{1}$$

where $F'(x) = f(x)$ and c is an arbitrary constant. Of the many formulas given in typical tables of integrals there are only a few that appear as a matter of course in typical texts in DEs. If $u = g(x)$ is a differentiable function, then its differential is defined as $du = g'(x)dx$. Each of the following results can be verified by differentiation.

$$\int u^n du = \frac{u^{n+1}}{n+1} + c, \quad n \neq 1 \tag{2}$$

$$\int u^{-1} du = \int \frac{1}{u} du = \ln|u| + c \tag{3}$$

$$\int e^u du = e^u + c \tag{4}$$

$$\int \sin u \, du = -\cos u + c \tag{5}$$

$$\int \cos u \, du = \sin u + c \tag{6}$$

$$\int \sec^2 u \, du = \tan u + c \tag{7}$$

$$\int \csc^2 u \, du = -\cot u + c \tag{8}$$

$$\int \frac{1}{a^2 + u^2} \, du = \frac{1}{a} \tan^{-1}\left(\frac{u}{a}\right) + c \tag{9}$$

For example, to evaluate $\int \sqrt{5x - 2} \, dx$ we identify $u = 5x - 2$, $du = 5 \, dx$, and write

$$\int (5x - 2)^{1/2} \, dx = \frac{1}{5} \int \overbrace{(5x - 2)^{1/2}}^{u^{1/2}} \underbrace{5 \, dx}_{du} = \frac{1}{5}\frac{2}{3}(5x - 2)^{3/2} + c = \frac{2}{15}(5x - 2)^{3/2} + c.$$

You are encouraged to always check your answer. In the last example, in view of the Chain Rule we have

$$\frac{d}{dx}\left[\frac{2}{15}(5x - 2)^{3/2} + c\right] = \frac{2}{15}\frac{3}{2}(5x - 2)^{1/2} \cdot 5 = (5x - 2)^{1/2}.$$

A formula very similar to (9) uses inverse hyperbolic functions

$$\int \frac{1}{a^2 - u^2} \, du = \frac{1}{a} \tanh^{-1}\left(\frac{u}{a}\right) + c, \quad |u| < a. \tag{10}$$

Every inverse hyperbolic function is a natural logarithm. Hence, if the integrand on the left-hand side of (10) is expanded into partial fractions, then an alternative form of (10) is found to be

$$\int \frac{1}{a^2 - u^2} \, du = \frac{1}{2a} \ln\left|\frac{a + u}{a - u}\right| + c, \quad |u| \neq a. \tag{11}$$

Formulas (10) and (11) are useful in some of the problems in Exercises 3.2.

Integration by Parts Integration by parts is a method for integrating certain products of functions as well as some inverse functions such as $\ln x$ (the inverse of e^x) and $\sin^{-1} x$. If $u = f(x)$ and $v = g(x)$, then the differentials are $du = f'(x) \, dx$ and $dv = g'(x) \, dx$, respectively. Integration by parts is a two-step procedure that begins with an integration: Integrate the function that you choose to call dv, and then differentiate the other function (by default after dv is chosen, the function remaining is u):

$$\int u \,\boxed{dv} = \boxed{u}\,\boxed{v} - \int v \,\boxed{du}. \tag{12}$$

When choosing which portion of the integrand to designate dv you should take account of the fact that this function will need to be integrated, first to obtain v, and then integrated again to obtain

22

$\int v\,du$. Thus, you will normally want to choose dv both so that its integral is known to you and so that the integral of dv is no more complicated than v itself. For example, in $\int x^3 \sin x\,dx$ you want to let $dv = \sin x\,dx$ because $\sin x$ is easily integrated and $\int \sin x\,dx = -\cos x + c$ is no more complicated than $\sin x$. [If you chose $dv = x^3\,dx$, $\int x^3\,dx = \frac{1}{4}x^4 + c$ is more complicated than x^3.] There are exceptions however. For example, when finding $\int x^3 \ln x\,dx$, you should let $u = \ln x$ and $dv = x^3\,dx$.

Some Important Integrals   The following integrals require integration by parts.

(a) If $n = 1, 2, 3, \ldots$, the indefinite integral of the form $\int x^n e^{ax}\,dx$, a a constant, demands integration by parts n times. For example, to evaluate $\int x^2 e^{5x}\,dx$ we must use (12) twice. It is also important to remember that for integrals of this type, integration by parts always starts by integrating the exponential function. For example,

$$\int x\,\boxed{e^{-2x}}\,dx = \boxed{x}\,\boxed{\left(-\tfrac{1}{2}e^{-2x}\right)} - \int \boxed{1}\cdot\left(-\frac{1}{2}e^{-2x}\right)dx = -\frac{1}{2}xe^{-2x} - \frac{1}{4}e^{-2x} + c.$$

with "differentiate" bracket over x and $\left(-\tfrac{1}{2}e^{-2x}\right)$, and "integrate" bracket under e^{-2x} and x.

(b) An integral $\int x^\alpha \ln x\,dx$, $\alpha \neq -1$, begins by integrating x^α. For example,

$$\int x^3 \ln x\,dx = \frac{1}{4}x^4 \ln x - \frac{1}{4}\int x^4 \frac{1}{x}\,dx = \frac{1}{4}x^4 \ln x - \frac{1}{4}\int x^3\,dx = \frac{1}{4}x^4 \ln x - \frac{1}{16}x^4 + c.$$

The special case when $\alpha = -1$, does not require integration by parts. By writing

$$\int x^{-1}\ln x\,dx = \int \ln x\,\frac{dx}{x}$$

and identifying $u = \ln x$, $du = dx/x$, the integral is recognized as the form given in (2) with $n = 1$:

$$\int x^{-1}\ln x\,dx = \int \overbrace{\ln x}^{u^1}\,\overbrace{\frac{dx}{x}}^{du} = \frac{1}{2}(\ln x)^2 + c.$$

Verify this last result by differentiating the right-hand side.

(c) You might recall that integrals of the type $\int e^{\alpha x} \sin \beta x\,dx$ or $\int e^{\alpha x} \cos \beta x\,dx$ require integration by parts twice. The unusual feature of these integrals is that in the second application of integration by parts the original integral is recovered on the right-hand side. We then solve for the original integral. It is hard to make a mistake here since we can begin the process by integrating either the exponential function or the trigonometric function. But it is important to remember that if the first integraton by parts is begun by integrating, say, the exponential function, then the second integration by parts must also begin by integrating the exponential function. For example,

$$\int \boxed{e^{2x}}\, \sin 3x\, dx = \boxed{\frac{1}{2}e^{2x}}\; \boxed{\sin 3x} - \int \frac{1}{2}e^{2x}\, \boxed{(3\cos 3x)}\, dx$$

$$= \frac{1}{2}e^{2x}\sin 3x - \frac{3}{2}\int \boxed{e^{2x}}\,\cos 3x\, dx = \frac{1}{2}e^{2x}\sin 3x - \frac{3}{2}\left[\boxed{\frac{1}{2}e^{2x}}\;\boxed{\cos 3x} - \int \frac{1}{2}e^{2x}\,\boxed{(-3\sin 3x)}\, dx\right]$$

$$= \frac{1}{2}e^{2x}\sin 3x - \frac{3}{4}e^{2x}\cos 3x - \frac{9}{4}\int e^{2x}\sin 3x\, dx.$$

If the original integral is denoted by the symbol I then the last result is equivalent to

$$I = \frac{1}{2}e^{2x}\sin 3x - \frac{3}{4}e^{2x}\cos 3x - \frac{9}{4}\,I.$$

Solving for I then gives $\frac{13}{4}I = \frac{1}{2}e^{2x}\sin 3x - \frac{3}{4}e^{2x}\cos 3x$ or

$$I = \int e^{2x}\sin 3x\, dx = \frac{2}{13}e^{2x}\sin 3x - \frac{3}{13}e^{2x}\cos 3x + c.$$

See **Use of Computers** later in this section for *Mathematica* and *Maple* implementations of both indefinite and definite integration.

Partial Fractions The following discussion illustrates a procedure for integrating certain rational functions $P(x)/Q(x)$, where we shall assume for this discussion that the degree of $P(x)$ is less than the degree of $Q(x)$. (If the degree $P(x) \geq$ degree $Q(x)$, then start the process with long division.) This method, known as **partial fractions**, consists of expanding such a rational function into simpler component fractions, and then evaluating the integral term-by-term. We begin with an example. When the terms in the sum

$$\frac{2}{x+5} + \frac{1}{x+1} \tag{13}$$

are combined by means of a common denominator, we obtain the single rational expression

$$\frac{3x+7}{(x+5)(x+1)}. \tag{14}$$

Now suppose that we are faced with the problem of evaluating the integral $\displaystyle\int \frac{3x+7}{(x+5)(x+1)}\, dx$. Of course, the solution is obvious. We use the equality of (13) and (14) to write:

$$\int \frac{3x+7}{(x+5)(x+1)}\, dx = \int \frac{2}{x+5}\, dx + \int \frac{1}{x+1}\, dx = 2\ln|x+5| + \ln|x+1| + c.$$

In the discussion that follows, we review the assumptions and the algebra of four cases of partial fraction decomposition.

CASE 1 Nonrepeated Linear Factors

We state the following fact from algebra without proof. If

$$\frac{P(x)}{Q(x)} = \frac{P(x)}{(a_1x + b_1)(a_2x + b_2) \cdots (a_nx + b_n)}$$

where all the factors $a_ix + b_i$, $i = 2, \ldots, n$, in the denominator are distinct and the degree of $P(x)$ is less than n, then unique real constants $C_1, C_2, \ldots, C_n$ exist such that

$$\frac{P(x)}{Q(x)} = \frac{C_1}{a_1x + b_1} + \frac{C_2}{a_2x + b_2} + \cdots + \frac{C_n}{a_nx + b_n}.$$

For example, to expand $(2x + 1)/(x - 1)(x + 3)$ into individual partial fractions we make the assumption that the integrand can be written as

$$\frac{2x + 1}{(x - 1)(x + 3)} = \frac{A}{x - 1} + \frac{B}{x + 3}.$$

Combining the terms of the right-hand member of the equation over a common denominator gives

$$\frac{2x + 1}{(x - 1)(x + 3)} = \frac{A(x + 3) + B(x - 1)}{(x - 1)(x + 3)}.$$

Since the denominators are identical, the numerators are also identical

$$2x + 1 = A(x + 3) + B(x - 1) = (A + B)x + (3A - B), \tag{15}$$

and the coefficients of the powers of x are the same:

$$2 = A + B \quad \text{and} \quad 1 = 3A - B.$$

These simultaneous equations can now be solved for A and B. The results are $A = \frac{3}{4}$ and $B = \frac{5}{4}$. Therefore,

$$\frac{2x + 1}{(x - 1)(x + 3)} = \frac{\frac{3}{4}}{x - 1} + \frac{\frac{5}{4}}{x + 3}.$$

Note: The numbers A and B in the preceding example can be determined in an alternative manner. Since (15) is an identity, that is, the equality is true for every value of x, it holds for $x = 1$ and $x = -3$, *the zeros of the denominator*. Setting $x = 1$ in (15) gives $3 = 4A$, from which it follows immediately that $A = \frac{3}{4}$. Similarly, by setting $x = -3$ in (15), we obtain $-5 = (-4)B$ or $B = \frac{5}{4}$. In the *Remarks* at the end of Section 7.2 in the text an equivalent and simpler method for evaluating the coefficients in a partial fraction decomposition is given. It is important to be aware that this technique, called the **cover-up method**, works only when the denomintor is a product of linear factors, none of which are repeated.

CASE II   Repeated Linear Factors

If

$$\frac{P(x)}{Q(x)} = \frac{P(x)}{(ax+b)^n}$$

where $n > 1$ and the degree of $P(x)$ is less than n, then unique real constants $C_1, C_2, \ldots, C_n$ can be found such that

$$\frac{P(x)}{(ax+b)^n} = \frac{C_1}{ax+b} + \frac{C_2}{(ax+b)^2} + \cdots + \frac{C_n}{(ax+b)^n}.$$

Combining the Cases: When the denominator $Q(x)$ contains distinct as well as repeated inear factors, we combine the two cases. For example, to expand $(6x-1)/x^3(2x-1)$ into partial fractions we write

$$\frac{6x-1}{x^3(2x-1)} = \frac{A}{x} + \frac{B}{x^2} + \frac{C}{x^3} + \frac{D}{2x-1}$$

since x^3 means that the linear term x is used as a factor x three times. Putting the right-hand side over a common denominator and equating numerators gives

$$6x - 1 = Ax^2(2x-1) + Bx(2x-1) + C(2x-1) + Dx^3 \tag{16}$$

$$= (2A+D)x^3 + (-A+2B)x^2 + (-B+2C)x - C. \tag{17}$$

If we set $x = 0$ and $x = \frac{1}{2}$ in (16), we find $C = 1$ and $D = 16$, respectively. Now since the denominator of the original expression has only two distinct zeros, we can find A and B by equating the coefficients of x^3 and x^2 in

$$6x - 1 = 0x^3 + 0x^2 + 6x - 1 = (2A+16)x^3 + (-A+2B)x^2 + (-B+2)x - 1.$$

Thus, $A = -8$ and $B = -4$. Therefore,

$$\frac{6x-1}{x^3(2x-1)} = -\frac{8}{x} - \frac{4}{x^2} + \frac{1}{x^3} + \frac{16}{2x-1}.$$

CASE III   Nonrepeated Quadratic Factors

Suppose the denominator of the rational function $P(x)/Q(x)$ can be expressed as a product of distinct quadratic factors $a_i x^2 + b_i x + c_i$, $i = 1, 2, \ldots, n$, that are irreducible (not factorable) over the real numbers. If the degree of $P(x)$ is less than $2n$, we can find unique real constants $A_1, A_2, \ldots, A_n, B_1, B_2, \ldots, B_n$ such that

$$\frac{P(x)}{(a_1 x^2 + b_1 x + c_1)(a_2 x^2 + b_2 x + c_2) \cdots (a_n x^2 + b_n x + c_n)}$$

$$= \frac{A_1 x + B_1}{a_1 x^2 + b_1 x + c} + \frac{A_2 x + B_2}{a_2 x^2 + b_2 x + c_2} + \cdots + \frac{A_n x + B_n}{a_n x^2 + b_n x + c_n}.$$

For example, to expand $4x/(x^2+1)(x^2+2x+3)$ into partial fractions we first write

$$\frac{4x}{(x^2+1)(x^2+2x+3)} = \frac{Ax+B}{x^2+1} + \frac{Cx+D}{x^2+2x+3}$$

and, after putting the right-hand side over a common denominator, equate numerators:

$$4x = (Ax+B)(x^2+2x+3) + (Cx+D)(x^2+1)$$

$$= (A+B)x^3 + (2A+B+D)x^2 + (3A+2B+C)x + (3B+D).$$

Since the denominator of the original fraction has no real zeros, we compare coefficients of powers of x:

$$0 = A+C$$

$$0 = 2A+B+D$$

$$4 = 3A+2B+C$$

$$0 = 3B+D.$$

Solving the equations yields $A=1$, $B=1$, $C=-1$, and $D=-3$. Therefore,

$$\frac{4x}{(x^2+1)(x^2+2x+3)} = \frac{x+1}{x^2+1} - \frac{x+3}{x^2+2x+3}.$$

CASE IV Repeated Quadratic Factors

In the case when the denominator contains a repeated irreducible quadratic polynomial ax^2+bx+c,

$$\frac{P(x)}{Q(x)} = \frac{P(x)}{(ax^2+bx+c)^n},$$

then if $P(x)$ is of degree less than $2n$, we can find real constants $A_1, A_2, \ldots, A_n, B_1, B_2, \ldots, B_n$ such that

$$\frac{P(x)}{(ax^2+bx+c)^n} = \frac{A_1x+B_1}{ax^2+bx+c} + \frac{A_2x+B_2}{(ax^2+bx+c)^2} + \cdots + \frac{A_nx+B_n}{(ax^2+bx+c)^n}.$$

Combining the Cases: The denominator $Q(x)$ in the fraction $(x+3)/x^2(x-5)(x^2+1)^2$ contains a distinct linear factor $x-5$, repeated linear factors x^2, and repeated irreducible quadratic factors $(x^2+1)^2$. To expand into partial fractions, we write

$$\frac{x+3}{x^2(x-5)(x^2+1)^2} = \underbrace{\frac{A}{x} + \frac{B}{x^2}}_{repeated\ linear} + \underbrace{\frac{C}{x-5}}_{linear} + \underbrace{\frac{Dx+E}{x^2+1} + \frac{Fx+G}{(x^2+1)^2}}_{repeated\ quadratic}.$$

Important Note: Although we have used it as a technique of integration, the ability to expand a rational expression $P(x)/Q(x)$ into partial fractions is also used in another context in the study of DEs; *it is frequently essential* when computing the inverse Laplace transform in Chapter 7. See

Section 7.2 in this manual for the syntax used in *Mathematica* and *Maple* to obtain a partial fraction decomposition of a rational function.

Use of Computers   We show below how *Mathematica* and *Maple* can be used to graph a function, graph an equation, find (and name) the solution of an equation, and find the indefinite integral (or antiderivative) and definite integral of a function.

Graphing a Function: To graph a function, such as $f(x) = \sqrt{4 - x^2}$, use the syntax

```
Clear[f]                                          (Mathematica)
f[x_]:= Sqrt[4 - x^2]
f[x]
Plot[f[x], {x, -3, 3}]

f:=x->sqrt(4-x^2);                                (Maple)
plot(f(x), x = -3..3);
```

Because $\sqrt{4 - x^2}$ is not a real number for $x < -2$ or $x > 2$, *Mathematica* returns some error messages when it tries to evaluate the function at points in these intervals; it does however then plot the graph. *Maple*, on the other hand, appears to ignore such problems and simply plots the graph. Since the graph of $f(x)$ is the upper half of the circle $x^2 + y^2 = 4$, you would expect the graph to look like a semicircle. Due, however, to the fact that both programs choose the widths and heights of their plots without regard to the scales on the horizontal and vertical axes, the semicircle may look more like a semi-ellipse. To insure that one unit on each axis has the same length on the computer screen, a graphics option is used as shown below.

```
Plot[f[x], {x, -2, 2}, AspectRatio->2/4]          (Mathematica)
plot(f(x), x=-3..3, scaling=CONSTRAINED);          (Maple)
```

(The **AspectRatio** option in the **Plot** command is the height of the plot divided by its width.)

Both programs provide a large number of graphics options that can be used to control the appearance of plots. Probably the most commonly used of these is an option that specifies the extent of the vertical axis to be shown. This may be necessary to see details of the graph that are obscured if the CAS (as opposed to the user) is allowed to choose the range of the graph shown. Examples are shown below with pictures displayed for the *Mathematica* output. The pictures generated by *Maple* are similar.

```
Plot[(x^2 - 1)Exp[2x^2], {x, -2, 2},              (Mathematica)
    PlotRange->{-2, 2}]
```

```
plot((x^2-1)*exp(2*x^2), x = -2..2,        (Maple)
    y = -2..2));
```

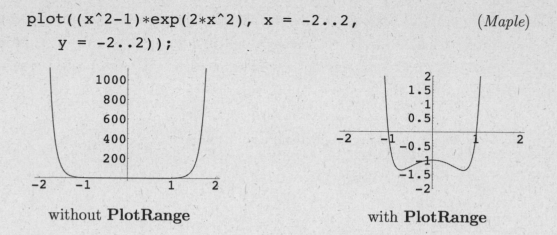

without **PlotRange**   with **PlotRange**

Finally, you may want to superimpose the graphs of two or more functions on the same set of coordinate axes. The commands below show the graphs of $f(x) = x^3 - 3x^2 + x + 1$ and $g(x) = 1 - 5x/4$, the tangent line to the graph of $f(x)$ at $x = 3/2$.

> **Clear[f, g]**   (*Mathematica*)
> **f[x_]:= x^3 - 3x^2 + x + 1**
> **g[x_]:= 1 - 5x/4**
> **{f[x], g[x]}**
> **Plot[{f(x), g(x)}, {x, -2, 3}]**

```
f:=x->x^3-3*x^2+x+1;                        (Maple)
g:=x->1-5*x/4;
plot({f(x),g(x)}, x = -2..3);
```

Graphing an Equation: In addition to graphing functions, a CAS can be used to graph equations that implicitly define functions. Equations such as this frequently arise when solving a DE using separation of variables. For example, the solution of $(1 + e^y)y' = x \cos x + \sin x$, $y(0) = 0$, obtained using separation of variables, is $y + e^y = x \sin x + 1$. We cannot solve this equation for y in terms of elementary functions, but we can graph the equation using **ContourPlot** in *Mathematica* and **contourplot** in *Maple* as shown below.

> **f[x_, y_]:= y + Exp[y] - x Sin[x] - 1**   (*Mathematica*)
> **f[x, y]**
> **ContourPlot[f[x, y], {x, -15, 15}, {y, -10, 5}, PlotPoints->50,**
> **Contours->{0}, ContourShading->False, Frame->False,**
> **Axes->True, AxesOrigin->{0, 0}, AxesLabel->{"x", "y"}]**

29

Many of the graphics options used above, such as **Frame** and **Axes** are self-explanatory. The default setting for **PlotPoints** is 15 and may result in a graph consisting of straight line segments and sharp corners. The **Contours** options is set to 0 in this case because the equation being graphed is $f(x,y) = y + e^y - x \sin x - 1 = 0$. Alternatively, the function could have been specified as $f(x,y) = y + e^x - x \sin x$ and the **Contours** option set to 1.

In *Maple* we effect the same result using the commands

```
f:=(x,y)->y+exp(y)-x*sin(x)-1;                    (Maple)
with(plots);
contourplot(f(x,y), x=-15..15, y=-10..5,
   contours=[0], grid=[80,80]);
```

The `grid` option is set to numbers that are large enough to insure a fairly smooth-looking graph.

Solving an Equation: Depending on the nature of an equation, there are a number of commands in both *Mathematica* and *Maple* that can be used to find solutions of the equation. If the equation involves only polynomial or rational functions, use **Solve** or **NSolve** in *Mathematica* and `solve` in *Maple*.

$$\text{Solve}[2x^2 - 3x - 4 == 0, x] \qquad (Mathematica)$$

or

$$\text{NSolve}[2x^2 - 3x - 4 == 0, x]$$

and

```
solve(2*x^2-3*x-4=0, x);                          (Maple)
```

Both **Solve** and `solve` will give exact solutions when possible. To obtain decimal versions of the solutions use **NSolve** in *Mathematica* and insert a decimal point in one of the numbers in the equation (`2*x^2-3*x-4.=0`) in *Maple*. (This will also work with the **Solve** command in *Mathematica* as an alternative to **NSolve**.)

To solve equations involving more complicated expressions such as $e^x = 3\cos x + 1$ use **FindRoot** in *Mathematica* and `fsolve` in *Maple*. In both programs, either Newton's method or a variation of it is used, so some form of an initial guess should be provided. This can be obtained from graphs and will enable you to select the solution you want in the event there is more than one.

$$\text{Plot}[\{\text{Exp}[x], 3\text{Cos}[x]+1\}, \{x, -5, 5\}] \qquad (Mathematica)$$
$$\text{FindRoot}[\text{Exp}[x]==3\text{Cos}[x]+1, \{x, 1\}]$$

30

```
plot([exp(x), 3*cos(x)+1], x=-5..5, y=-3..10);   (Maple)
fsolve(exp(x), 3*cos(x)+1, x=0..2);
```

As shown above, in *Maple*, a range of values can be specified. Other options are available, but this is generally the most effective way to find the solution you are looking for. If there is no solution within this range, no solution will be given. If there are multiple solutions within the range, only one will be given, unless you are solving a polynomial equation, in which case all solutions within the range will be given.

Naming the Solution of an Equation: If you intend to use the solution of an equation in a subsequent computation in your CAS, then it is convenient to have a name for this value so that you do not have to read it from the screen and then type it in. The routines below show how to do this, assigning the name a to the solution of $e^{3x-x^2}\cos x = 0$.

```
Clear[f]                                          (Mathematica)
f[x_]:= Exp[x - x^2] + Cos[x]
f[x]
Plot[f[x], {x, -5, 5}]
sol = FindRoot[f[x]== 0, {x, 2}]
a = x/.sol
```

```
f:=x->exp(x-x^2)+cos(x);                          (Maple)
plot(f(x), x=-5..5);
a:=fsolve(f(x)=0, x=1..3);
```

Integrals: *Mathematica* and *Maple* can find both antiderivatives (indefinite integrals) of many functions and definite integrals of most functions. To find an antiderivative of $f(x)$ use

Integrate[f[x], x]   (Mathematica)

```
int(f(x), x);                                     (Maple)
```

If you want to further work with the antiderivative in the CAS it can be helpful to define it as a function that can itself be evaluated at a point, graphed, or differentiated. The syntax to accomplish this is

g[x_]:= Evaluate[Integrate[f(x), x]]   (Mathematica)
g[x]

```
g:=unapply(int(f(x),x),x);                        (Maple)
```

31

To find the definite integral, $\int_a^b f(x)\,dx$, of f we use

$$\textbf{Integrate[f(x), \{x, a, b\}]} \hspace{4cm} (\textit{Mathematica})$$

$$\texttt{int(f(x),x=a..b);} \hspace{4cm} (\textit{Maple})$$

The comands **Integrate** and `int` will try to find exact values of a definite integral by evaluating the antiderivative at the endpoints of the interval. If the CAS is unable to find an antiderivative of $f(x)$ (for example, when $f(x) = \sin(\sin x)$, or if you simply want a numerical approximation of the definite integral, use

$$\textbf{NIntegrate[f(x), \{x, a, b\}]} \hspace{4cm} (\textit{Mathematica})$$

$$\texttt{evalf(int(f(x),x=a..b));} \hspace{4cm} (\textit{Maple})$$

Exercises 2.2	*Hints, Suggestions, Solutions, and Examples*

In many of the following problems we will encounter an expression of the form $\ln|g(y)| = f(x) + c$. To solve for $g(y)$ we exponentiate both sides of the equation. This yields $|g(y)| = e^{f(x)+c} = e^c e^{f(x)}$ which implies $g(y) = \pm e^c e^{f(x)}$. Letting $c_1 = \pm e^c$ we obtain $g(y) = c_1 e^{f(x)}$.

3. From $dy = -e^{-3x}\,dx$ we obtain $y = \frac{1}{3}e^{-3x} + c$.

6. From $\dfrac{1}{y^2}\,dy = -2x\,dx$ we obtain $-\dfrac{1}{y} = -x^2 + c$ or $y = \dfrac{1}{x^2 + c_1}$.

9. From $\left(y + 2 + \dfrac{1}{y}\right)dy = x^2 \ln x\,dx$ we obtain $\dfrac{y^2}{2} + 2y + \ln|y| = \dfrac{x^3}{3}\ln|x| - \dfrac{1}{9}x^3 + c$.

12. From $2y\,dy = -\dfrac{\sin 3x}{\cos^3 3x}\,dx$ or $2y\,dy = -\tan 3x\sec^2 3x\,dx$ we obtain $y^2 = -\frac{1}{6}\sec^2 3x + c$.

15. From $\dfrac{1}{S}\,dS = k\,dr$ we obtain $S = ce^{kr}$.

18. From $\dfrac{1}{N}\,dN = \left(te^{t+2} - 1\right)dt$ we obtain $\ln|N| = te^{t+2} - e^{t+2} - t + c$ or $N = c_1 e^{te^{t+2} - e^{t+2} - t}$.

21. From $x\,dx = \dfrac{1}{\sqrt{1 - y^2}}\,dy$ we obtain $\frac{1}{2}x^2 = \sin^{-1}y + c$ or $y = \sin\left(\dfrac{x^2}{2} + c_1\right)$.

24. From $\dfrac{1}{y^2 - 1}\,dy = \dfrac{1}{x^2 - 1}\,dx$ or $\dfrac{1}{2}\left(\dfrac{1}{y - 1} - \dfrac{1}{y + 1}\right)dy = \dfrac{1}{2}\left(\dfrac{1}{x - 1} - \dfrac{1}{x + 1}\right)dx$ we obtain

$\ln|y-1| - \ln|y+1| = \ln|x-1| - \ln|x+1| + \ln c$ or $\dfrac{y-1}{y+1} = \dfrac{c(x-1)}{x+1}$. Using $y(2) = 2$ we find

$c = 1$. A solution of the IVP is $\dfrac{y-1}{y+1} = \dfrac{x-1}{x+1}$ or $y = x$.

27. Separating variables and integrating we obtain

$$\frac{dx}{\sqrt{1-x^2}} - \frac{dy}{\sqrt{1-y^2}} = 0 \quad \text{and} \quad \sin^{-1} x - \sin^{-1} y = c.$$

Setting $x = 0$ and $y = \sqrt{3}/2$ we obtain $c = -\pi/3$. Thus, an implicit solution of the IVP is $\sin^{-1} x - \sin^{-1} y = -\pi/3$. Solving for y and using an addition formula from trigonometry, we get

$$y = \sin\left(\sin^{-1} x + \frac{\pi}{3}\right) = x \cos\frac{\pi}{3} + \sqrt{1-x^2}\,\sin\frac{\pi}{3} = \frac{x}{2} + \frac{\sqrt{3}\sqrt{1-x^2}}{2}.$$

30. Separating variables, integrating from -2 to x, and using t as a dummy variable of integration gives

$$\int_{-2}^{x} \frac{1}{y^2}\frac{dy}{dt}\,dt = \int_{-2}^{x} \sin t^2 dt$$

$$-y(t)^{-1}\Big|_{-2}^{x} = \int_{-2}^{x} \sin t^2 dt$$

$$-y(x)^{-1} + y(-2)^{-1} = \int_{-2}^{x} \sin t^2 dt$$

$$-y(x)^{-1} = -y(-2)^{-1} + \int_{-2}^{x} \sin t^2 dt$$

$$y(x)^{-1} = 3 - \int_{-2}^{x} \sin t^2 dt.$$

Thus

$$y(x) = \frac{1}{3 - \int_{-2}^{x} \sin t^2 dt}.$$

33. Singular solutions of $dy/dx = x\sqrt{1-y^2}$ are $y = -1$ and $y = 1$. A singular solution of $(e^x + e^{-x})dy/dx = y^2$ is $y = 0$.

36. Separating variables we obtain $\dfrac{dy}{(y-1)^2} = dx.$ Then

$$-\frac{1}{y-1} = x+c \quad \text{and} \quad y = \frac{x+c-1}{x+c}.$$

Setting $x = 0$ and $y = 1.01$ we obtain $c = -100$. The solution is

$$y = \frac{x-101}{x-100}.$$

39. Separating variables, we have

$$\frac{dy}{y - y^3} = \frac{dy}{y(1-y)(1+y)} = \left(\frac{1}{y} + \frac{1/2}{1-y} - \frac{1/2}{1+y}\right) dy = dx.$$

Integrating, we get

$$\ln|y| - \frac{1}{2}\ln|1-y| - \frac{1}{2}\ln|1+y| = x+c.$$

When $y > 1$, this becomes

$$\ln y - \frac{1}{2}\ln(y-1) - \frac{1}{2}\ln(y+1) = \ln\frac{y}{\sqrt{y^2-1}} = x+c.$$

Letting $x = 0$ and $y = 2$ we find $c = \ln(2/\sqrt{3})$. Solving for y we get $y_1(x) = 2e^x/\sqrt{4e^{2x} - 3}$, where $x > \ln(\sqrt{3}/2)$.

When $0 < y < 1$ we have

$$\ln y - \frac{1}{2}\ln(1-y) - \frac{1}{2}\ln(1+y) = \ln\frac{y}{\sqrt{1-y^2}} = x+c.$$

Letting $x = 0$ and $y = \frac{1}{2}$ we find $c = \ln(1/\sqrt{3})$. Solving for y we get $y_2(x) = e^x/\sqrt{e^{2x} + 3}$, where $-\infty < x < \infty$.

When $-1 < y < 0$ we have

$$\ln(-y) - \frac{1}{2}\ln(1-y) - \frac{1}{2}\ln(1+y) = \ln\frac{-y}{\sqrt{1-y^2}} = x+c.$$

Letting $x = 0$ and $y = -\frac{1}{2}$ we find $c = \ln(1/\sqrt{3})$. Solving for y we get $y_3(x) = -e^x/\sqrt{e^{2x} + 3}$, where $-\infty < x < \infty$.

When $y < -1$ we have

$$\ln(-y) - \frac{1}{2}\ln(1-y) - \frac{1}{2}\ln(-1-y) = \ln\frac{-y}{\sqrt{y^2-1}} = x+c.$$

Letting $x = 0$ and $y = -2$ we find $c = \ln(2/\sqrt{3}\,)$. Solving for y we get $y_4(x) = -2e^x/\sqrt{4e^{2x} - 3}$, where $x > \ln(\sqrt{3}/2)$.

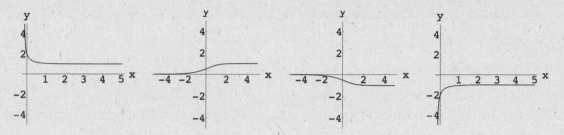

42. (a) From Problem 7 the general solution is $3e^{-2y} + 2e^{3x} = c$. When $y(0) = 0$ we find $c = 5$, so $3e^{-2y} + 2e^{3x} = 5$. Solving for y we get $y = -\frac{1}{2}\ln\frac{1}{3}(5 - 2e^{3x})$.

(b) The interval of definition appears to be approximately $(-\infty, 0.3)$.

(c) Solving $\frac{1}{3}(5 - 2e^{3x}) = 0$ we get $x = \frac{1}{3}\ln(\frac{5}{2})$, so the exact interval of definition is $(-\infty, \frac{1}{3}\ln\frac{5}{2})$.

50. (*Suggestion*) First, explicitly solve the DE and then use **ContourPlot** in *Mathematica* or `contourplot` in *Maple* as described earlier in this section of this manual to obtain level curves. In either CAS the contours option can be set to a list of values for the constant.

52. (*Suggestion*) See the suggestion above for Problem 50.

Section 2.3 Linear Equations

The terminology and concepts listed below provide an outline of the main ideas encountered in this section. These can be useful when preparing for a quiz or test.

Terminology and Concepts

- linear first-order DE

- homogeneous linear DE

- variation of parameters as applied to a linear first-order DE

- integrating factor

- general solution of a linear first-order DE

- singular points of a linear first-order DE

- transient term

- special functions

The basic skills listed below summarize the more mechanical types of problems encountered in the exercise set for this section.

Basic Skills

- find the integrating factor for a linear first-order DE

- use the integrating factor to find the general solution of a linear first-order DE

- find transient terms in the general solution of a linear first-order DE

- solve a linear first-order DE satisfying an initial condition

- solve a linear first-order IVP with a piecewise-defined coefficient

- determine an interval of definition of the general solution of a linear first-order DE

An Identity   In precalculus mathematics you learned that the exponential statement $e^N = M$ is equivalent to the logarithmic expression $N = \ln M$. If we substitute this last expression into the first, we obtain the important identity

$$e^{\ln M} = M,$$

which is very useful in this section when computing the integrating factor $e^{\int P(x)\,dx}$. For example, to determine the integrating factor for $(5+x^2)\dfrac{dy}{dx} - 2xy = x$ we first put the DE into standard form $\dfrac{dy}{dx} - \dfrac{2x}{5+x^2}\,y = \dfrac{x}{5+x^2}$ and identify $P(x) = -2x/(5+x^2)$:

$$e^{-\int \frac{2x}{5+x^2}\,dx} = e^{-\ln|5+x^2|} = e^{\ln|5+x^2|^{-1}} = (5+x^2)^{-1} = \frac{1}{5+x^2}.$$

We were able to discard the absolute value sign in $|5+x^2|$ because $5+x^2 > 0$ for all x.

It may take some time to get used to the fact that once the DE has been put into standard form and the integrating factor determined, multiplying both sides of the standard form by the integrating factor recasts the left-hand side of the equation into the derivative of the integrating factor and the dependent variable. Above, we saw that the integrating factor for $\dfrac{dy}{dx} - \dfrac{2x}{5+x^2}\,y = \dfrac{x}{5+x^2}$ is $\dfrac{1}{5+x^2}$. Multiplying the DE by this factor gives

$$\frac{1}{5+x^2}\frac{dy}{dx} - \frac{2x}{(5+x^2)^2}\,y = \frac{x}{(5+x^2)^2}. \tag{1}$$

The left-hand side of the last DE can then be written compactly as the derivative of a product of the integrating factor and $y(x)$:

$$\frac{d}{dx}\left[\frac{1}{5+x^2}y\right] = \frac{x}{(5+x^2)^2}.\qquad(2)$$

Until you become comfortable with this notion we suggest that you actually multiply the differential equation $dy/dx + P(x)y = f(x)$ by the integrating factor and then verify using the product rule that

$$e^{\int P(x)\,dx}\frac{dy}{dx} + P(x)e^{\int P(x)\,dx}y = e^{\int P(x)\,dx}f(x)\qquad\text{(as in (1))}$$

and

$$\frac{d}{dx}\left[e^{\int P(x)\,dx}y\right] = e^{\int P(x)\,dx}f(x)\qquad\text{(as in (2))}$$

are formally equivalent. After that, once you have computed the integrating factor, you can jump to the last form and proceed with integration. In the case of (2) the integral of the left-hand side is

$$\int\frac{d}{dx}\left[\frac{1}{5+x^2}y\right]dx = \frac{1}{5+x^2}y.$$

The integral of the right-hand side is

$$\int\frac{x}{(5+x^2)^2}\,dx = \frac{1}{2}\int(5+x^2)^{-2}(2x\,dx) = -\frac{1}{2}(5+x^2)^{-1}+c.$$

Hence the solution of the DE on $(-\infty,\infty)$ is

$$\frac{1}{5+x^2}y = -\frac{1}{2}(5+x^2)^{-1}+c\qquad\text{or}\qquad y = -\frac{1}{2}+c(5+x^2).$$

Use of Computers As we have seen in Section 2.2, a CAS can help you find the solution of a DE by performing some of the more complicated integrations required. In addition to this, a CAS can also frequently find directly the solution of a DE. For example, to solve the DE $xy' - xy = e^x$ use

```
Dsolve[x y'[x] - x y[x]==Exp[x], y[x], x]          (Mathematica)

dsolve(x*diff(y(x),x)-x*y(x)=exp(x),y(x));         (Maple)
```

To solve an IVP such as $xy' - xy = e^x$, $y(1) = 2$, use

```
Dsolve[{x y'[x] - x y[x]==Exp[x], y[1]==2}, y[x], x]     (Mathematica)

dsolve({x*diff(y(x),x)-x*y(x)=exp(x),y(1)=2},       (Maple)
   y(x));
```

The output is

$$\{\{y[x] \to e^{x-1}(2 + e \ln x)\}\} \qquad\qquad (Mathematica)$$

$$y[x] = \left(\ln x + \frac{2}{e}\right) e^x \qquad\qquad (Maple)$$

In neither case can the output be used subsequently in the CAS to easily obtain numerical values or graphs of the function which is the solution, $y(x)$, of the IVP. The following routines can be used to assign a specific function name to the solution. This function can then be evaluated at a number, differentiated, graphed, etc.

Clear[y] $\qquad\qquad\qquad\qquad\qquad\qquad\qquad$ (*Mathematica*)
sol=Flatten[Dsolve[{x^2 y'[x] - x y[x]==y[x]^2, y[1]==2}, y[x], x]];
y[x_]:=Evaluate[y[x]/.sol]
y[x]

The semicolon at the end of line of input in *Mathematica* is quite different than in *Maple*, where it is required at the end of every line of input. In *Mathematica* it simply suppresses the display of the output. For example, typing **a=2** in *Mathematica* will set a equal to 2 and display 2 on the monitor, whereas typing **a=2;** will set a equal to 2, but will not display the 2.

```
sol:=dsolve({x^2*diff(y(x),x)-x*y(x)=y(x)^2,     (Maple)
    y(1)=2}, y(x));
y:=unapply(eval(y(x),sol), x);
```

Special Functions: As discussed in the text, a number of nonelementary functions are defined by integrals. Others, such as the Bessel functions discussed in Chapter 6 of the text, can be defined using infinite series. Virtually all special functions that have been tabulated in standard handbooks are built into *Mathematica* and *Maple*. Below is a table listing all of the special functions used in the text, together with the syntax for each function in both *Mathematica* and *Maple*.

Name	Notation	*Mathematica*	*Maple*
Bessel function of the first kind	$J_n(x)$	**BesselJ[n,x]**	`BesselJ(n,x)`
Bessel function of the second kind	$Y_n(x)$	**BesselY[n,x]**	`BesselY(n,x)`
Chebyshev polynomial	$T_n(x)$	**ChebyshevT[n,x]**	`ChebyshevT(n,x)`
Complementary error function	$\text{erfc}(x)$	**Erfc[x]**	`erfc(x)`

Dirac delta function	$\delta(t-a)$	**DiracDelta[t-a]**	`Dirac(t-a)`
Error function	$\text{erf}(x)$	**Erf[x]**	`erf(x)`
Fresnel sine integral	$S(x)$	**FresnelS[x]**	`FresnelS(x)`
Gamma	$\Gamma(x)$	**Gamma[x]**	`GAMMA(x)`
Heaviside (unit step function)	$(t-a)$	**UnitStep[t-a]**	`Heaviside(x)`
Hermite polynomial	$H_n(x)$	**HermiteH[n,x]**	`HermiteH(n,x)`
Imaginary error function	$\text{erfi}(x)$	**Erfi[x]**	`erfi(x)`
Laguerre polynomial	$L_n(x)$	**LaguerreL[n,x]**	`orthopoly[L](n,x)`
Legendre polynomial	$P_n(x)$	**LegendreP[n,x]**	`orthopoly[P](n,x)`
Modified Bessel function of the first kind	$I(x)$	**BesselI[n,x]**	`BesselI(n,x)`
Modified Bessel function of the second kind	$K(x)$	**BesselK[n,x]**	`BesselK(n,x)`
Sine integral	$Si(x)$	**SinIntegral[x]**	`Si(x)`

Exercises 2.3   *Hints, Suggestions, Solutions, and Examples*

See Section 4.6 in this manual for a brief discussion of the integrals of $\tan x$, $\cot x$, $\sec x$, *and* $\csc x$. *These will be used in several of the problems below.*

3. For $y' + y = e^{3x}$ an integrating factor is $e^{\int dx} = e^x$ so that $\dfrac{d}{dx}\left[e^x y\right] = e^{4x}$ and $y = \frac{1}{4}e^{3x} + ce^{-x}$ for $-\infty < x < \infty$. The transient term is ce^{-x}.

6. For $y' + 2xy = x^3$ an integrating factor is $e^{\int 2x\,dx} = e^{x^2}$ so that $\dfrac{d}{dx}\left[e^{x^2} y\right] = x^3 e^{x^2}$ and $y = \frac{1}{2}x^2 - \frac{1}{2} + ce^{-x^2}$ for $-\infty < x < \infty$. The transient term is ce^{-x^2}.

9. For $y' - \dfrac{1}{x}y = x\sin x$ an integrating factor is $e^{-\int (1/x)dx} = \dfrac{1}{x}$ so that $\dfrac{d}{dx}\left[\dfrac{1}{x}y\right] = \sin x$ and $y = cx - x\cos x$ for $0 < x < \infty$. There is no transient term.

12. For $y' - \dfrac{x}{(1+x)} y = x$ an integrating factor is $e^{-\int [x/(1+x)]dx} = (x+1)e^{-x}$ so that $\dfrac{d}{dx}\left[(x+1)e^{-x}y\right] = x(x+1)e^{-x}$ and $y = -x - \dfrac{2x+3}{x+1} + \dfrac{ce^x}{x+1}$ for $-1 < x < \infty$. There is no transient term.

14. (*Hint*) To compute the integrating factor, $e^{\int [(1+x)/x]dx}$, begin by writing

$$\frac{1+x}{x} = \frac{1}{x} + 1.$$

15. For $\dfrac{dx}{dy} - \dfrac{4}{y}x = 4y^5$ an integrating factor is $e^{-\int(4/y)dy} = e^{\ln y^{-4}} = y^{-4}$ so that $\dfrac{d}{dy}\left[y^{-4}x\right] = 4y$ and $x = 2y^6 + cy^4$ for $0 < y < \infty$. There is no transient term.

18. For $y' + (\cot x)y = \sec^2 x \csc x$ an integrating factor is $e^{\int \cot x \, dx} = e^{\ln |\sin x|} = \sin x$ so that $\dfrac{d}{dx}\left[(\sin x)\,y\right] = \sec^2 x$ and $y = \sec x + c\csc x$ for $0 < x < \pi/2$. There is no transient term.

19. (*Hint*) To compute the integrating factor, $e^{\int [(x+2)/(x+1)]dx}$, begin by writing

$$\frac{x+2}{x+1} = \frac{x+1+1}{x+1} = 1 + \frac{1}{x+1},$$

or obtain this result using long division.

21. For $\dfrac{dr}{d\theta} + r\sec\theta = \cos\theta$ an integrating factor is $e^{\int \sec\theta \, d\theta} = e^{\ln|\sec x+\tan x|} = \sec\theta + \tan\theta$ so that $\dfrac{d}{d\theta}\left[(\sec\theta + \tan\theta)r\right] = 1 + \sin\theta$ and $(\sec\theta + \tan\theta)r = \theta - \cos\theta + c$ for $-\pi/2 < \theta < \pi/2$.

24. For $y' + \dfrac{2}{x^2 - 1}y = \dfrac{x+1}{x-1}$ an integrating factor is $e^{\int[2/(x^2-1)]dx} = \dfrac{x-1}{x+1}$ so that $\dfrac{d}{dx}\left[\dfrac{x-1}{x+1}y\right] = 1$ and $(x-1)y = x(x+1) + c(x+1)$ for $-1 < x < 1$.

27. For $\dfrac{di}{dt} + \dfrac{R}{L}i = \dfrac{E}{L}$ an integrating factor is $e^{\int (R/L)\,dt} = e^{Rt/L}$ so that $\dfrac{d}{dt}\left[e^{Rt/L}\,i\right] = \dfrac{E}{L}e^{Rt/L}$ and $i = \dfrac{E}{R} + ce^{-Rt/L}$ for $-\infty < t < \infty$. If $i(0) = i_0$ then $c = i_0 - E/R$ and $i = \dfrac{E}{R} + \left(i_0 - \dfrac{E}{R}\right)e^{-Rt/L}$.

30. For $y' + (\tan x)y = \cos^2 x$ an integrating factor is $e^{\int \tan x \, dx} = e^{\ln |\sec x|} = \sec x$ so that $\dfrac{d}{dx}\left[(\sec x)\,y\right] = \cos x$ and $y = \sin x \cos x + c\cos x$ for $-\pi/2 < x < \pi/2$. If $y(0) = -1$ then $c = -1$ and $y = \sin x \cos x - \cos x$.

31-34. (*Suggestion*) See **Use of Computers** in Section 1.1 of this manual for the syntax used in a CAS to specify piecewise-defined functions.

33. For $y' + 2xy = f(x)$ an integrating factor is e^{x^2} so that

$$ye^{x^2} = \begin{cases} \frac{1}{2}e^{x^2} + c_1, & 0 \le x \le 1 \\ c_2, & x > 1. \end{cases}$$

If $y(0) = 2$ then $c_1 = 3/2$ and for continuity we must have $c_2 = \frac{1}{2}e + \frac{3}{2}$ so that

$$y = \begin{cases} \frac{1}{2} + \frac{3}{2}e^{-x^2}, & 0 \le x \le 1 \\ \left(\frac{1}{2}e + \frac{3}{2}\right)e^{-x^2}, & x > 1. \end{cases}$$

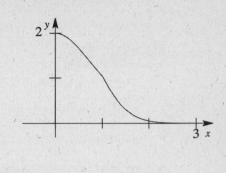

36. For $y' + e^x y = 1$ an integrating factor is e^{e^x}. Thus

$$\frac{d}{dx}\left[e^{e^x} y\right] = e^{e^x} \quad \text{and} \quad e^{e^x} y = \int_0^x e^{e^t} dt + c.$$

From $y(0) = 1$ we get $c = e$, so $y = e^{-e^x} \int_0^x e^{e^t} dt + e^{1-e^x}$.

When $y' + e^x y = 0$ we can separate variables and integrate:

$$\frac{dy}{y} = -e^x \, dx \quad \text{and} \quad \ln|y| = -e^x + c.$$

Thus $y = c_1 e^{-e^x}$. From $y(0) = 1$ we get $c_1 = e$, so $y = e^{1-e^x}$.

When $y' + e^x y = e^x$ we can see by inspection that $y = 1$ is a solution.

Section 2.4 Exact Equations

The terminology and concepts listed below provide an outline of the main ideas encountered in this section. These can be useful when preparing for a quiz or test.

Terminology and Concepts

- differential of a function f of two variables

- exact DE

- criterion for exactness of $M(x, y)\, dx + N(x, y)\, dy = 0$

- use of an integrating factor to change a nonexact DE into one that is exact

The basic skills listed below summarize the more mechanical types of problems encountered in the exercise set for this section.

Basic Skills

- determine whether a DE is exact

- solve an exact DE

- find and use an integrating factor to make a DE exact

Partial Differentiation In this section you will have to use **partial differentiation** and a reverse process which could be called **partial integration** or **partial antidifferentiation.**

Suppose $f(x, y)$ is a function of two variables defined in a region R of the plane. The mechanics of computing the partial derivative of f with respect to x, written $\partial f/\partial x$, can be summarized simply as:

To compute $\dfrac{\partial f}{\partial x}$, use the rules of ordinary differentiation with respect to x while treating y as a constant.

For example, if $f(x, y) = 3x^4 y^2 + e^{5y} + 4\cos xy$, then by treating y as a constant we obtain

$$\frac{\partial f}{\partial x} = y^2 \frac{\partial}{\partial x} 3x^4 + \frac{\partial}{\partial x} e^{5y} + 4 \frac{\partial}{\partial x} \cos xy^2 = y^2 \cdot 12x^3 + 0 + 4(-y^2 \sin xy^2)$$

$$= 12x^3 y^2 - 4y^2 \sin xy^2.$$

On the other hand, to compute the partial derivative of f with respect to y, written $\partial f/\partial y$, use the rules of ordinary differentiation, except treat x as a constant. For example, for the function f in the preceding example, we have

$$\frac{\partial f}{\partial y} = 3x^4 \frac{\partial}{\partial y} y^2 + \frac{\partial}{\partial y} e^{5y} + 4 \frac{\partial}{\partial y} \cos xy^2 = 3x^4 \cdot 2y + 5e^{5y} + 4(-x \cdot 2y \sin xy^2)$$

$$= 6x^4 y + 5e^{5y} - 8xy \sin xy^2.$$

Now suppose that we are given the partial derivative of a function f, say $\partial f/\partial x = M(x, y)$. We can recover f up to an "additive constant" by partial integration with respect to x:

$$f(x, y) = \int M(x, y)\, dx.$$

In the foregoing integral we use the rules of ordinary indefinite integration with respect to x while treating y as a constant. If $\partial f/\partial y = N(x, y)$, then by partial integration with respect to y we have

$$f(x, y) = \int N(x, y)\, dy,$$

where x is treated as a constant. For example, if $\partial f/\partial x = x + y$, then

$$f(x, y) = \int (x + y)\, dx = \frac{1}{2}x^2 + xy + g(y).$$

42

Here any function $g(y)$ is a "constant" of integration since

$$\frac{\partial f}{\partial x} = \frac{1}{2}\frac{\partial}{\partial x}x^2 + y\frac{\partial}{\partial x}x + \frac{\partial}{\partial x}g(y) = \frac{1}{2}\cdot 2x + y\cdot 1 + 0 = x + y.$$

As another example, suppose that $\partial f/\partial y = 10e^{-x^2 y}$, then, with x held constant,

$$f(x,y) = \int 10e^{-x^2 y}dy = \frac{10}{-x^2}\int e^{-x^2 y}(-x^2\,dy) = -\frac{10}{x^2}e^{-x^2 y} + h(x),$$

where this time the "constant" of integration is $h(x)$. You are encouraged to verify the last result by partial differentiation with respect to y.

Use of Computers To obtain the partial derivative of $f(x,y) = xe^y + x^2y^3$ with respect to y in a CAS use

```
Clear[f]                              (Mathematica)
f[x_, y_]:= x Exp[y] + x^2 y^3
f[x, y]
D[f[x, y], y]

f:=(x,y)->x*exp(y)+x^2*y^3;           (Maple)
diff(f(x,y),y);    or    D[2](f);
```

In *Maple*, `diff(f(x,y),y);` returns the expression $xe^y + 3x^2y^2$, which cannot be directly evaluated at a point. On the other hand, `D[2](f);` returns the function $(x,y) \rightarrow xe^y + 3x^2y^2$. In this case, the 2 in `D[2]` signifies the partial derivative with respect to the second variable, y, in the definition of f.

Exercises 2.4   *Hints, Suggestions, Solutions, and Examples*

3. Let $M = 5x + 4y$ and $N = 4x - 8y^3$ so that $M_y = 4 = N_x$. From $f_x = 5x + 4y$ we obtain $f = \frac{5}{2}x^2 + 4xy + h(y)$, $h'(y) = -8y^3$, and $h(y) = -2y^4$. A solution is $\frac{5}{2}x^2 + 4xy - 2y^4 = c$.

6. Let $M = 4x^3 - 3y\sin 3x - y/x^2$ and $N = 2y - 1/x + \cos 3x$ so that $M_y = -3\sin 3x - 1/x^2$ and $N_x = 1/x^2 - 3\sin 3x$. The equation is not exact.

9. Let $M = y^3 - y^2\sin x - x$ and $N = 3xy^2 + 2y\cos x$ so that $M_y = 3y^2 - 2y\sin x = N_x$. From $f_x = y^3 - y^2\sin x - x$ we obtain $f = xy^3 + y^2\cos x - \frac{1}{2}x^2 + h(y)$, $h'(y) = 0$, and $h(y) = 0$. A solution is $xy^3 + y^2\cos x - \frac{1}{2}x^2 = c$.

12. Let $M = 3x^2y + e^y$ and $N = x^3 + xe^y - 2y$ so that $M_y = 3x^2 + e^y = N_x$. From $f_x = 3x^2y + e^y$ we obtain $f = x^3y + xe^y + h(y)$, $h'(y) = -2y$, and $h(y) = -y^2$. A solution is $x^3y + xe^y - y^2 = c$.

15. Let $M = x^2y^3 - 1/\left(1 + 9x^2\right)$ and $N = x^3y^2$ so that $M_y = 3x^2y^2 = N_x$. From $f_x = x^2y^3 - 1/\left(1 + 9x^2\right)$ we obtain $f = \frac{1}{3}x^3y^3 - \frac{1}{3}\arctan(3x) + h(y)$, $h'(y) = 0$, and $h(y) = 0$. A solution is $x^3y^3 - \arctan(3x) = c$.

18. Let $M = 2y\sin x \cos x - y + 2y^2 e^{xy^2}$ and $N = -x + \sin^2 x + 4xy e^{xy^2}$ so that

$$M_y = 2\sin x \cos x - 1 + 4xy^3 e^{xy^2} + 4y e^{xy^2} = N_x.$$

From $f_x = 2y\sin x \cos x - y + 2y^2 e^{xy^2}$ we obtain $f = y\sin^2 x - xy + 2e^{xy^2} + h(y)$, $h'(y) = 0$, and $h(y) = 0$. A solution is $y\sin^2 x - xy + 2e^{xy^2} = c$.

21. Let $M = x^2 + 2xy + y^2$ and $N = 2xy + x^2 - 1$ so that $M_y = 2(x+y) = N_x$. From $f_x = x^2 + 2xy + y^2$ we obtain $f = \frac{1}{3}x^3 + x^2y + xy^2 + h(y)$, $h'(y) = -1$, and $h(y) = -y$. The solution is $\frac{1}{3}x^3 + x^2y + xy^2 - y = c$. If $y(1) = 1$ then $c = 4/3$ and a solution of the IVP is $\frac{1}{3}x^3 + x^2y + xy^2 - y = \frac{4}{3}$.

24. Let $M = t/2y^4$ and $N = \left(3y^2 - t^2\right)/y^5$ so that $M_y = -2t/y^5 = N_t$. From $f_t = t/2y^4$ we obtain

$$f = \frac{t^2}{4y^4} + h(y), \quad h'(y) = \frac{3}{y^3}, \quad \text{and } h(y) = -\frac{3}{2y^2}. \quad \text{The solution is } \frac{t^2}{4y^4} - \frac{3}{2y^2} = c. \quad \text{If } y(1) = 1 \text{ then}$$

$c = -5/4$ and a solution of the IVP is $\dfrac{t^2}{4y^4} - \dfrac{3}{2y^2} = -\dfrac{5}{4}$.

27. Equating $M_y = 3y^2 + 4kxy^3$ and $N_x = 3y^2 + 40xy^3$ we obtain $k = 10$.

30. Let $M = \left(x^2 + 2xy - y^2\right)/\left(x^2 + 2xy + y^2\right)$ and $N = \left(y^2 + 2xy - x^2\right)/\left(y^2 + 2xy + x^2\right)$ so that $M_y = -4xy/(x+y)^3 = N_x$. From $f_x = \left(x^2 + 2xy + y^2 - 2y^2\right)/(x+y)^2$ we obtain

$$f = x + \frac{2y^2}{x+y} + h(y), \quad h'(y) = -1, \quad \text{and } h(y) = -y. \text{ A solution of the DE is } x^2 + y^2 = c(x+y).$$

33. We note that $(N_x - M_y)/M = 2/y$, so an integrating factor is $e^{\int 2dy/y} = y^2$. Let $M = 6xy^3$ and $N = 4y^3 + 9x^2y^2$ so that $M_y = 18xy^2 = N_x$. From $f_x = 6xy^3$ we obtain $f = 3x^2y^3 + h(y)$, $h'(y) = 4y^3$, and $h(y) = y^4$. A solution of the DE is $3x^2y^3 + y^4 = c$.

36. We note that $(N_x - M_y)/M = -3/y$, so an integrating factor is $e^{-3\int dy/y} = 1/y^3$. Let

$$M = (y^2 + xy^3)/y^3 = 1/y + x$$

and

$$N = (5y^2 - xy + y^3\sin y)/y^3 = 5/y - x/y^2 + \sin y,$$

so that $M_y = -1/y^2 = N_x$. From $f_x = 1/y + x$ we obtain $f = x/y + \frac{1}{2}x^2 + h(y)$, $h'(y) = 5/y + \sin y$, and $h(y) = 5\ln|y| - \cos y$. A solution of the DE is $x/y + \frac{1}{2}x^2 + 5\ln|y| - \cos y = c$.

39. (a) Implicitly differentiating $x^3 + 2x^2y + y^2 = c$ and solving for dy/dx we obtain

$$3x^2 + 2x^2\frac{dy}{dx} + 4xy + 2y\frac{dy}{dx} = 0 \quad \text{and} \quad \frac{dy}{dx} = -\frac{3x^2 + 4xy}{2x^2 + 2y}.$$

By writing the last equation in differential form we get $(4xy + 3x^2)dx + (2y + 2x^2)dy = 0$.

44

(b) Setting $x = 0$ and $y = -2$ in $x^3 + 2x^2y + y^2 = c$ we find $c = 4$, and setting $x = y = 1$ we also find $c = 4$. Thus, both initial conditions determine the same implicit solution.

(c) Solving $x^3 + 2x^2y + y^2 = 4$ for y we get

$$y_1(x) = -x^2 - \sqrt{4 - x^3 + x^4}$$

and

$$y_2(x) = -x^2 + \sqrt{4 - x^3 + x^4}.$$

Observe in the figure that $y_1(0) = -2$ and $y_2(1) = 1$.

44. (*Hint*) Rewrite the DE as $-g(x)\,dx + (1/h(y))\,dy = 0$.

46. (*Suggestion*) See the discussion in this section of the *SRSM* for a way to graph several functions in a CAS.

Section 2.5

Solutions by Substitutions

The terminology and concepts listed below provide an outline of the main ideas encountered in this section. These can be useful when preparing for a quiz or test.

Terminology and Concepts

- homogeneous DE
- Bernoulli's equation
- use of a substitution to reduce a DE of the form $dy/dx = f(Ax + By + C)$ to a separable DE

The basic skills listed below summarize the more mechanical types of problems encountered in the exercise set for this section.

Basic Skills

- recognize and solve a homogeneous DE using an appropriate substitution
- recognize a Bernoulli equation and solve it by reducing it to a linear first-order DE
- solve a DE of the form $dy/dx = f(Ax + By + C)$ using an appropriate substitution

Homogeneous DEs   As mentioned in the text, either substitution $y = ux$ or $x = vy$ will reduce a homogeneous DE $M(x, y)\,dx + N(x, y)\,dy = 0$ to one with separable variables. As a rule of thumb, try the substitution $y = ux$ whenever the coefficient N is simpler than M. On the other hand, try the substitution $x = vy$ whenever the coefficient M is simpler than N.

Exercises 2.5   *Hints, Suggestions, Solutions, and Examples*

3. Letting $x = vy$ we have

$$vy(v\,dy + y\,dv) + (y - 2vy)\,dy = 0$$

$$vy^2\,dv + y\left(v^2 - 2v + 1\right)dy = 0$$

$$\frac{v\,dv}{(v-1)^2} + \frac{dy}{y} = 0$$

$$\ln|v-1| - \frac{1}{v-1} + \ln|y| = c$$

$$\ln\left|\frac{x}{y} - 1\right| - \frac{1}{x/y - 1} + \ln y = c$$

$$(x - y)\ln|x - y| - y = c(x - y).$$

6. Letting $y = ux$ and using partial fractions, we have

$$\left(u^2 x^2 + ux^2\right)dx + x^2(u\,dx + x\,du) = 0$$

$$x^2\left(u^2 + 2u\right)dx + x^3\,du = 0$$

$$\frac{dx}{x} + \frac{du}{u(u+2)} = 0$$

$$\ln|x| + \frac{1}{2}\ln|u| - \frac{1}{2}\ln|u+2| = c$$

$$\frac{x^2 u}{u+2} = c_1$$

$$x^2\frac{y}{x} = c_1\left(\frac{y}{x} + 2\right)$$

$$x^2 y = c_1(y + 2x).$$

9. Letting $y = ux$ we have

$$-ux\,dx + (x + \sqrt{u}\,x)(u\,dx + x\,du) = 0$$

$$(x^2 + x^2\sqrt{u})\,du + xu^{3/2}\,dx = 0$$

$$\left(u^{-3/2} + \frac{1}{u}\right)du + \frac{dx}{x} = 0$$

$$-2u^{-1/2} + \ln|u| + \ln|x| = c$$

$$\ln|y/x| + \ln|x| = 2\sqrt{x/y} + c$$

$$y(\ln|y| - c)^2 = 4x.$$

12. Letting $y = ux$ we have

$$(x^2 + 2u^2x^2)dx - ux^2(u\,dx + x\,du) = 0$$

$$x^2(1 + u^2)dx - ux^3\,du = 0$$

$$\frac{dx}{x} - \frac{u\,du}{1 + u^2} = 0$$

$$\ln|x| - \frac{1}{2}\ln(1 + u^2) = c$$

$$\frac{x^2}{1 + u^2} = c_1$$

$$x^4 = c_1(x^2 + y^2).$$

Using $y(-1) = 1$ we find $c_1 = 1/2$. The solution of the IVP is $2x^4 = y^2 + x^2$.

15. From $y' + \frac{1}{x}y = \frac{1}{x}y^{-2}$ and $w = y^3$ we obtain $\frac{dw}{dx} + \frac{3}{x}w = \frac{3}{x}$. An integrating factor is x^3 so that $x^3w = x^3 + c$ or $y^3 = 1 + cx^{-3}$.

18. From $y' - \left(1 + \frac{1}{x}\right)y = y^2$ and $w = y^{-1}$ we obtain $\frac{dw}{dx} + \left(1 + \frac{1}{x}\right)w = -1$. An integrating factor is xe^x so that $xe^xw = -xe^x + e^x + c$ or $y^{-1} = -1 + \frac{1}{x} + \frac{c}{x}e^{-x}$.

21. From $y' - \frac{2}{x}y = \frac{3}{x^2}y^4$ and $w = y^{-3}$ we obtain $\frac{dw}{dx} + \frac{6}{x}w = -\frac{9}{x^2}$. An integrating factor is x^6 so that $x^6w = -\frac{9}{5}x^5 + c$ or $y^{-3} = -\frac{9}{5}x^{-1} + cx^{-6}$. If $y(1) = \frac{1}{2}$ then $c = \frac{49}{5}$ and $y^{-3} = -\frac{9}{5}x^{-1} + \frac{49}{5}x^{-6}$.

24. Let $u = x + y$ so that $du/dx = 1 + dy/dx$. Then $\frac{du}{dx} - 1 = \frac{1 - u}{u}$ or $u\,du = dx$. Thus $\frac{1}{2}u^2 = x + c$ or $u^2 = 2x + c_1$, and $(x + y)^2 = 2x + c_1$.

27. Let $u = y - 2x + 3$ so that $du/dx = dy/dx - 2$. Then $\dfrac{du}{dx} + 2 = 2 + \sqrt{u}$ or $\dfrac{1}{\sqrt{u}}\,du = dx$. Thus $2\sqrt{u} = x + c$ and $2\sqrt{y - 2x + 3} = x + c$.

30. Let $u = 3x + 2y$ so that $du/dx = 3 + 2\,dy/dx$. Then $\dfrac{du}{dx} = 3 + \dfrac{2u}{u+2} = \dfrac{5u+6}{u+2}$ and $\dfrac{u+2}{5u+6}\,du = dx$. Now by long division

$$\frac{u+2}{5u+6} = \frac{1}{5} + \frac{4}{25u+30}$$

so we have

$$\int \left(\frac{1}{5} + \frac{4}{25u+30} \right) du = dx$$

and $\frac{1}{5}u + \frac{4}{25}\ln|25u + 30| = x + c$. Thus

$$\frac{1}{5}(3x + 2y) + \frac{4}{25}\ln|75x + 50y + 30| = x + c.$$

Setting $x = -1$ and $y = -1$ we obtain $c = \frac{4}{25}\ln 95$. The solution is

$$\frac{1}{5}(3x + 2y) + \frac{4}{25}\ln|75x + 50y + 30| = x + \frac{4}{25}\ln 95$$

or

$$5y - 5x + 2\ln|75x + 50y + 30| = 2\ln 95.$$

33. (*Hint*) To find a singular solution, try functions $y(x)$ that have a simplifying effect on the DE.

Section 2.6 A Numerical Method

The terminology and concepts listed below provide an outline of the main ideas encountered in this section. These can be useful when preparing for a quiz or test.

Terminology and Concepts

- direction field
- linearization of a function at a point
- Euler's method
- absolute error
- percentage relative error
- numerical solver

The basic skills listed below summarize the more mechanical types of problems encountered in the exercise set for this section.

Basic Skills

- use Euler's method to approximate the solution of a first-order DE

Use of Computers

Both *Mathematica* and *Maple* contain sophisticated programming languages. Below are examples of how each CAS can be used to implement Euler's method and display a table similar to the ones shown in Example 2 in this section of the text. Here we will consider the IVP $y' = f(x, y) = 0.2xy$, $y(1) = 1$, and use Euler's method to approximate $y(1.5)$ using the step size $h = 0.1$. To compute the absolute error at each step we will use the fact that the analytic solution is $y = g(x) = e^{0.1(x^2-1)}$.

```
Clear[f, g, u, v, t, err]                                    (Mathematica)
f[x_, y_]:= 0.2x y        (* Input line *)
Print["y'= ", f[x, y]]
g[x_]:= Exp[0.1(x^2 - 1)]        (* Input line *)
Print["y(x) = ", g[x]]
h = 0.1; a = 1; b = 1.5; y0 = 1;        (* Input line *)
u[0] = a; v[0] = y0;
u[n_]:= u[n] = u[n - 1] + h
v[n_]:= v[n] = v[n - 1] + h f[u[n - 1], v[n - 1]]
t[n_]:= t[n] = g[u[n]]
err[n_]:= err[n] = Abs[t[n] - v[n]]
approxtab = Table[{u[n], v[n], t[n], err[n]}, {n, 0, Round[(b - a)/h]}];
tableheader = {"x", "y", "actual value", "abs. error"};
Prepend[approxtab, tableheader]//TableForm
```

The use of the form $u[n\_]:= u[n] = u[n - 1] + h$ in the *Mathematica* code above is a way to define a function that remembers a value once it has been computed. This greatly speeds up the running of a program that involves a recursively defined function. To make it easier to reuse the routine for subsequent problems, the lines containing inputs to the routine can be commented, as shown above, or color coded.

```
f:=(x,y)->0.2*x*y;          #Input line          (Maple)
g:=x->exp(0.1*(x^2-1));       #Input line
h:=0.1; a:=1; b:=1.5; y0:=1;       #Input line
```

49

```
u:=n->u(n-1)+h; u(0):=a;
v:=n->v(n-1)+h*f(u(n-1),v(n-1)); v(0):=y0;
t:=n->g(u(n));
err:=n->abs(t(n)-v(n));
u(-1):='x'; v(-1):='y'; t(-1):='actual_value';
err(-1):='abs._error';
with(linalg)
matrix(2+round((b-a)/h),4,(i,j)->piecewise(j=1,u(i-2),
    j=2,v(i-2),j=3,t(i-2),j=4,err(i-2)));
```

Exercises 2.6   *Hints, Suggestions, Solutions, and Examples*

3. Separating variables and integrating, we have

$$\frac{dy}{y} = dx \quad \text{and} \quad \ln|y| = x + c.$$

Thus $y = c_1 e^x$ and, using $y(0) = 1$, we find $c = 1$, so $y = e^x$ is the solution of the IVP.

$h=0.1$

x_n	y_n	Actual Value	Abs. Error	% Rel. Error
0.00	1.0000	1.0000	0.0000	0.00
0.10	1.1000	1.1052	0.0052	0.47
0.20	1.2100	1.2214	0.0114	0.93
0.30	1.3310	1.3499	0.0189	1.40
0.40	1.4641	1.4918	0.0277	1.86
0.50	1.6105	1.6487	0.0382	2.32
0.60	1.7716	1.8221	0.0506	2.77
0.70	1.9487	2.0138	0.0650	3.23
0.80	2.1436	2.2255	0.0820	3.68
0.90	2.3579	2.4596	0.1017	4.13
1.00	2.5937	2.7183	0.1245	4.58

$h=0.05$

x_n	y_n	Actual Value	Abs. Error	% Rel. Error
0.00	1.0000	1.0000	0.0000	0.00
0.05	1.0500	1.0513	0.0013	0.12
0.10	1.1025	1.1052	0.0027	0.24
0.15	1.1576	1.1618	0.0042	0.36
0.20	1.2155	1.2214	0.0059	0.48
0.25	1.2763	1.2840	0.0077	0.60
0.30	1.3401	1.3499	0.0098	0.72
0.35	1.4071	1.4191	0.0120	0.84
0.40	1.4775	1.4918	0.0144	0.96
0.45	1.5513	1.5683	0.0170	1.08
0.50	1.6289	1.6487	0.0198	1.20
0.55	1.7103	1.7333	0.0229	1.32
0.60	1.7959	1.8221	0.0263	1.44
0.65	1.8856	1.9155	0.0299	1.56
0.70	1.9799	2.0138	0.0338	1.68
0.75	2.0789	2.1170	0.0381	1.80
0.80	2.1829	2.2255	0.0427	1.92
0.85	2.2920	2.3396	0.0476	2.04
0.90	2.4066	2.4596	0.0530	2.15
0.95	2.5270	2.5857	0.0588	2.27
1.00	2.6533	2.7183	0.0650	2.39

6.

x_n	y_n
0.00	1.0000
0.10	1.1000
0.20	1.2220
0.30	1.3753
0.40	1.5735
0.50	1.8371

h=0.1

x_n	y_n
0.00	1.0000
0.05	1.0500
0.10	1.1053
0.15	1.1668
0.20	1.2360
0.25	1.3144
0.30	1.4039
0.35	1.5070
0.40	1.6267
0.45	1.7670
0.50	1.9332

h=0.05

9.

x_n	y_n
1.00	1.0000
1.10	1.0000
1.20	1.0191
1.30	1.0588
1.40	1.1231
1.50	1.2194

h=0.1

x_n	y_n
1.00	1.0000
1.05	1.0000
1.10	1.0049
1.15	1.0147
1.20	1.0298
1.25	1.0506
1.30	1.0775
1.35	1.1115
1.40	1.1538
1.45	1.2057
1.50	1.2696

h=0.05

12. Tables of values were computed using the Euler and RK4 methods. the resulting points were plotted and joined using **ListPlot** in *Mathematica*. A simplified version of the code used to do this is given in *Use of Computers* in this section.

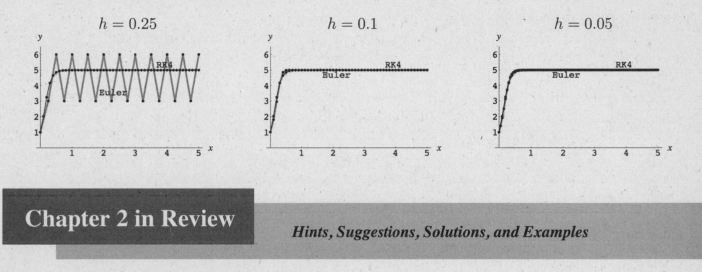

$h = 0.25$ $h = 0.1$ $h = 0.05$

Chapter 2 in Review

Hints, Suggestions, Solutions, and Examples

3. True; $y = k_2/k_1$ is always a solution for $k_1 \neq 0$.

5. (*Suggestion*) Read the solution below for Problem 6.

6. $\dfrac{dy}{dx} = y(y - 2)^2(y - 4)$

9.

12. Write the DE in the form

$$y \ln \frac{x}{y} \, dx = \left(x \ln \frac{x}{y} - y \right) dy.$$

This is a homogeneous equation, so let $x = uy$. Then $dx = u \, dy + y \, du$ and the DE becomes

$$y \ln u (u \, dy + y \, du) = (uy \ln u - y) \, dy \quad \text{or} \quad y \ln u \, du = -dy.$$

Separating variables, we obtain

$$\ln u \, du = -\frac{dy}{y}$$

$$u \ln |u| - u = -\ln |y| + c$$

$$\frac{x}{y} \ln \left| \frac{x}{y} \right| - \frac{x}{y} = -\ln |y| + c$$

$$x (\ln x - \ln y) - x = -y \ln |y| + cy.$$

15. Write the equation in the form

$$\frac{dQ}{dt} + \frac{1}{t} Q = t^3 \ln t.$$

An integrating factor is $e^{\ln t} = t$, so

$$\frac{d}{dt} [tQ] = t^4 \ln t$$

$$tQ = -\frac{1}{25} t^5 + \frac{1}{5} t^5 \ln t + c$$

and

$$Q = -\frac{1}{25} t^4 + \frac{1}{5} t^4 \ln t + \frac{c}{t}.$$

18. Letting $M = 2r^2 \cos \theta \sin \theta + r \cos \theta$ and $N = 4r + \sin \theta - 2r \cos^2 \theta$ we see that

$$M_r = 4r \cos \theta \sin \theta + \cos \theta = N_\theta,$$

so the DE is exact. From $f_\theta = 2r^2 \cos \theta \sin \theta + r \cos \theta$ we obtain $f = -r^2 \cos^2 \theta + r \sin \theta + h(r)$. Then

$$f_r = -2r \cos^2 \theta + \sin \theta + h'(r) = 4r + \sin \theta - 2r \cos^2 \theta$$

and $h'(r) = 4r$ so $h(r) = 2r^2$. The solution is

$$-r^2 \cos^2 \theta + r \sin \theta + 2r^2 = c.$$

21. (a) For $y < 0$, $\sqrt{y}$ is not a real number.

(b) Separating variables and integrating we have

$$\frac{dy}{\sqrt{y}} = dx \quad \text{and} \quad 2\sqrt{y} = x + c.$$

Letting $y(x_0) = y_0$ we get $c = 2\sqrt{y_0} - x_0$, so that

$$2\sqrt{y} = x + 2\sqrt{y_0} - x_0 \quad \text{and} \quad y = \frac{1}{4}(x + 2\sqrt{y_0} - x_0)^2.$$

Since $\sqrt{y} > 0$ for $y \neq 0$, we see that $dy/dx = \frac{1}{2}(x + 2\sqrt{y_0} - x_0)$ must be positive. Thus, the interval on which the solution is defined is $(x_0 - 2\sqrt{y_0}, \infty)$.

24. The first step of Euler's method gives $y(1.1) \approx 9 + 0.1(1+3) = 9.4$. Applying Euler's method one more time gives $y(1.2) \approx 9.4 + 0.1(1 + 1.1\sqrt{9.4}) \approx 9.8373$.

3 Modeling with First-Order Differential Equations

Linear Models

The terminology and concepts listed below provide an outline of the main ideas encountered in this section. These can be useful when preparing for a quiz or test.

Terminology and Concepts

- growth and decay
- growth and decay constant
- half-life
- carbon dating
- Newton's law of cooling/warming
- mixtures
- series circuits
- steady-state part of a solution
- transient term in solution

The basic skills listed below summarize the more mechanical types of problems encountered in the exercise set for this section.

Basic Skills

- set up and solve DEs modeling populations that exhibit exponential growth or decay
- set up and solve DEs involving Newton's law of cooling/warming
- set up and solve DEs modeling mixtures of fluids
- set up and solve DEs modeling series circuits
- set up and solve DEs modeling falling objects both with and without air resistance

Half-life   To drive the point home that "the usual carbon-14 technique is limited to about nine half-lives of the isotope," we will estimate the amount of C-14 that would remain in a piece of wood from a tree about 50,000 years after the tree is cut down. We use the fact that the half-life of C-14 is approximately 5,600 years. Then, in $2(5,600) = 12,200$ years, $\frac{1}{4}$ of the original amount of C-14 remains; in $3(5,600) = 16,800$ years, $\frac{1}{8}$ of the original amount remains. Continuing in this fashion, in $9(5,600) = 50,400$ years, $\frac{1}{512}$ (or only about 0.2%) of the original amount of C-14 remains.

Exercises 3.1   *Hints, Suggestions, Solutions, and Examples*

2-3. (*Hint*) Remember that some questions about how fast a solution of a DE changes can be answered by simply considering the DE, that is, without explicitly solving it.

3. Let $P = P(t)$ be the population at time t. Then $dP/dt = kP$ and $P = ce^{kt}$. From $P(0) = c = 500$ we see that $P = 500e^{kt}$. Since 15% of 500 is 75, we have $P(10) = 500e^{10k} = 575$. Solving for k, we get $k = \frac{1}{10}\ln\frac{575}{500} = \frac{1}{10}\ln 1.15$. When $t = 30$,

$$P(30) = 500e^{(1/10)(\ln 1.15)30} = 500e^{3\ln 1.15} = 760 \text{ years}$$

and

$$P'(30) = kP(30) = \frac{1}{10}(\ln 1.15)760 = 10.62 \text{ persons/year.}$$

6. Let $A = A(t)$ be the amount present at time t. From $dA/dt = kA$ and $A(0) = 100$ we obtain $A = 100e^{kt}$. Using $A(6) = 97$ we find $k = \frac{1}{6}\ln 0.97$. Then $A(24) = 100e^{(1/6)(\ln 0.97)24} = 100(0.97)^4 \approx 88.5$ mg.

9. Let $I = I(t)$ be the intensity, t the thickness, and $I(0) = I_0$. If $dI/dt = kI$ and $I(3) = 0.25I_0$, then $I = I_0e^{kt}$, $k = \frac{1}{3}\ln 0.25$, and $I(15) = 0.00098I_0$.

12. From Example 3 in the text, the amount of carbon present at time t is $A(t) = A_0e^{-0.00012378t}$. Letting $t = 660$ and solving for A_0 we have $A(660) = A_0e^{-0.0001237(660)} = 0.921553A_0$. Thus, approximately 92% of the original amount of C-14 remained in the cloth as of 1988.

15. We use the fact that the boiling temperature for water is $100°$ C. Now assume that $dT/dt = k(T - 100)$ so that $T = 100 + ce^{kt}$. If $T(0) = 20°$ and $T(1) = 22°$, then $c = -80$ and $k = \ln(39/40) \approx -0.0253$. Then $T(t) = 100 - 80e^{-0.0253t}$, and when $T = 90$, $t = 82.1$ seconds. If $T(t) = 98°$ then $t = 145.7$ seconds.

17. (*Hint*) The question posed at the end of this problem refers to the ambient temperature T_m of the oven.

18. **(a)** The initial temperature of the bath is $T_m(0) = 60°$, so in the short term the temperature of the chemical, which starts at $80°$, should decrease or cool. Over time, the temperature of the bath will increase toward $100°$ since $e^{-0.1t}$ decreases from 1 toward 0 as t increases from 0. Thus, in the long term, the temperature of the chemical should increase or warm toward $100°$.

(b) Adapting the model for Newton's law of cooling, we have

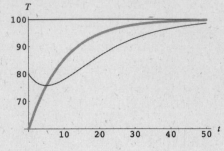

$$\frac{dT}{dt} = -0.1(T - 100 + 40e^{-0.1t}), \quad T(0) = 80.$$

Writing the DE in the form

$$\frac{dT}{dt} + 0.1T = 10 - 4e^{-0.1t}$$

we see that it is linear with integrating factor $e^{\int 0.1\,dt} = e^{0.1t}$. Thus

$$\frac{d}{dt}[e^{0.1t}T] = 10e^{0.1t} - 4$$

$$e^{0.1t}T = 100e^{0.1t} - 4t + c$$

and

$$T(t) = 100 - 4te^{-0.1t} + ce^{-0.1t}.$$

Now $T(0) = 80$ so $100 + c = 80$, $c = -20$ and

$$T(t) = 100 - 4te^{-0.1t} - 20e^{-0.1t} = 100 - (4t + 20)e^{-0.1t}.$$

The thinner curve verifies the prediction of cooling followed by warming toward $100°$. The wider curve shows the temperature T_m of the liquid bath.

21. From $dA/dt = 4 - A/50$ we obtain $A = 200 + ce^{-t/50}$. If $A(0) = 30$ then $c = -170$ and $A = 200 - 170e^{-t/50}$.

24. From Problem 23 the number of pounds of salt in the tank at time t is $A(t) = 1000 - 1000e^{-t/100}$. The concentration at time t is $c(t) = A(t)/500 = 2 - 2e^{-t/100}$. Therefore $c(5) = 2 - 2e^{-1/20} = 0.0975\,\text{lb/gal}$ and $\lim_{t \to \infty} c(t) = 2$. Solving $c(t) = 1 = 2 - 2e^{-t/100}$ for t we obtain $t = 100 \ln 2 \approx 69.3\,\text{min}$.

27. From

$$\frac{dA}{dt} = 3 - \frac{4A}{100 + (6 - 4)t} = 3 - \frac{2A}{50 + t}$$

we obtain $A = 50 + t + c(50 + t)^{-2}$. If $A(0) = 10$ then $c = -100{,}000$ and $A(30) = 64.38$ pounds.

30. Assume $L\,di/dt + Ri = E(t)$, $E(t) = E_0 \sin \omega t$, and $i(0) = i_0$ so that

$$i = \frac{E_0 R}{L^2 \omega^2 + R^2} \sin \omega t - \frac{E_0 L \omega}{L^2 \omega^2 + R^2} \cos \omega t + ce^{-Rt/L}.$$

Since $i(0) = i_0$ we obtain $c = i_0 + \dfrac{E_0 L \omega}{L^2 \omega^2 + R^2}$.

33. For $0 \leq t \leq 20$ the DE is $20\, di/dt + 2i = 120$. An integrating factor is $e^{t/10}$, so $(d/dt)[e^{t/10}i] = 6e^{t/10}$ and $i = 60 + c_1 e^{-t/10}$. If $i(0) = 0$ then $c_1 = -60$ and $i = 60 - 60e^{-t/10}$. For $t > 20$ the DE is $20\, di/dt + 2i = 0$ and $i = c_2 e^{-t/10}$. At $t = 20$ we want $c_2 e^{-2} = 60 - 60e^{-2}$ so that $c_2 = 60\left(e^2 - 1\right)$. Thus

$$i(t) = \begin{cases} 60 - 60e^{-t/10}, & 0 \leq t \leq 20 \\ 60\left(e^2 - 1\right)e^{-t/10}, & t > 20. \end{cases}$$

36. (a) Integrating $d^2s/dt^2 = -g$ we get $v(t) = ds/dt = -gt + c$. From $v(0) = 300$ we find $c = 300$, and we are given $g = 32$, so the velocity is $v(t) = -32t + 300$.

(b) Integrating again and using $s(0) = 0$ we get $s(t) = -16t^2 + 300t$. The maximum height is attained when $v = 0$, that is, at $t_a = 9.375$. The maximum height will be $s(9.375) = 1406.25$ ft.

39. (a) The DE is first-order and linear. Letting $b = k/\rho$, the integrating factor is $e^{\int 3b\, dt/(bt+r_0)} = (r_0 + bt)^3$. Then

$$\frac{d}{dt}[(r_0 + bt)^3 v] = g(r_0 + bt)^3 \quad \text{and} \quad (r_0 + bt)^3 v = \frac{g}{4b}(r_0 + bt)^4 + c.$$

The solution of the DE is $v(t) = (g/4b)(r_0 + bt) + c(r_0 + bt)^{-3}$. Using $v(0) = 0$ we find $c = -gr_0^4/4b$, so that

$$v(t) = \frac{g}{4b}(r_0 + bt) - \frac{gr_0^4}{4b(r_0 + bt)^3} = \frac{g\rho}{4k}\left(r_0 + \frac{k}{\rho}t\right) - \frac{g\rho r_0^4}{4k(r_0 + kt/\rho)^3}.$$

(b) Integrating $dr/dt = k/\rho$ we get $r = kt/\rho + c$. Using $r(0) = r_0$ we have $c = r_0$, so $r(t) = kt/\rho + r_0$.

(c) If $r = 0.007$ ft when $t = 10$ s, then solving $r(10) = 0.007$ for k/ρ, we obtain $k/\rho = -0.0003$ and $r(t) = 0.01 - 0.0003t$. Solving $r(t) = 0$ we get $t = 33.3$, so the raindrop will have evaporated completely at 33.3 seconds.

42. (a) The solution of the DE is $P(t) = c_1 e^{kt} + h/k$. If we let the initial population of fish be P_0 then $P(0) = P_0$ which implies that

$$c_1 = P_0 - \frac{h}{k} \quad \text{and} \quad P(t) = \left(P_0 - \frac{h}{k}\right)e^{kt} + \frac{h}{k}.$$

(b) For $P_0 > h/k$ all terms in the solution are positive. In this case $P(t)$ increases as time t increases. That is, $P(t) \to \infty$ as $t \to \infty$.

For $P_0 = h/k$ the population remains constant for all time t:

$$P(t) = \left(\frac{h}{k} - \frac{h}{k}\right)e^{kt} + \frac{h}{k} = \frac{h}{k}.$$

For $0 < P_0 < h/k$ the coefficient of the exponential function is negative and so the function decreases as time t increases.

(c) Since the function decreases and is concave down, the graph of $P(t)$ crosses the t-axis. That is, there exists a time $T > 0$ such that $P(T) = 0$. Solving

$$\left(P_0 - \frac{h}{k}\right)e^{kT} + \frac{h}{k} = 0$$

for T shows that the time of extinction is

$$T = \frac{1}{k}\ln\left(\frac{h}{h - kP_0}\right).$$

45. (a) For $0 \leq t < 4$, $6 \leq t < 10$ and $12 \leq t < 16$, no voltage is applied to the heart and $E(t) = 0$. At the other times, the differential equation is $dE/dt = -E/RC$. Separating variables, integrating, and solving for e, we get $E = ke^{-t/RC}$, subject to $E(4) = E(10) = E(16) = 12$. These intitial conditions yield, respectively, $k = 12e^{4/RC}$, $k = 12e^{10/RC}$, $k = 12e^{16/RC}$, and $k = 12e^{22/RC}$. Thus

$$E(t) = \begin{cases} 0, & 0 \leq t < 4, \ 6 \leq t < 10, \ 12 \leq t < 16 \\ 12e^{(4-t)/RC}, & 4 \leq t < 6 \\ 12e^{(10-t)/RC}, & 10 \leq t < 12 \\ 12e^{(16-t)/RC}, & 16 \leq t < 18 \\ 12e^{(22-t)/RC}, & 22 \leq t < 24. \end{cases}$$

(b)

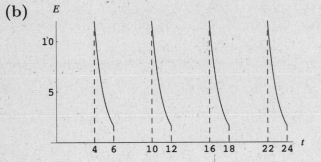

47. (*Suggestion*) In part (b) of this problem, rewrite the DE

$$m\frac{dv}{dt} = mg\sin\theta - \mu mg\cos\theta$$

as

$$m\frac{dv}{dt} = mg\cos\theta(\tan\theta - \mu).$$

Solve this DE subject to the initial condition that the box starts from rest in the two cases: $\tan\theta = \mu$ and $\tan\theta < \mu$.

48. (a) We saw in part (b) of Problem 36 of this section that the ascent time is $t_a = 9.375$. To find when the cannonball hits the ground we solve $s(t) = -16t^2 + 300t = 0$, getting a total time in flight of $t = 18.75\,\text{s}$. Thus, the time of descent is $t_d = 18.75 - 9.375 = 9.375$. The impact velocity is $v_i = v(18.75) = -300$, which has the same magnitude as the initial velocity.

(b) It is shown in Problem 37 of this section that the ascent time in the case of air resistance is $t_a = 9.162$. Solving $s(t) = 1{,}340{,}000 - 6{,}400t - 1{,}340{,}000e^{-0.005t} = 0$ we see that the total time of flight is 18.466 s. Thus, the descent time is $t_d = 18.466 - 9.162 = 9.304$. The impact velocity is $v_i = v(18.466) = -290.91$, compared to an initial velocity of $v_0 = 300$.

Section 3.2   Nonlinear Models

The terminology and concepts listed below provide an outline of the main ideas encountered in this section. These can be useful when preparing for a quiz or test.

Terminology and Concepts

- logistic DE
- logistic growth
- harvesting
- Gompertz DE
- second-order chemical reaction

The basic skills listed below summarize the more mechanical types of problems encountered in the exercise set for this section.

Basic Skills

- set up and solve DEs involving the logistic model
- set up and solve DEs involving chemical reactions
- set up and solve DEs involving falling bodies subject to air resistance that is proportional to the square of the instantaneous velocity

Use of Computers
A scatter plot in two dimensions is a set of points in the plane. Both *Mathematica* and *Maple* have commands for plotting a set of points. In the examples below we plot the points $(-1, -2.7)$, $(-0.5, -2.5)$, $(0, -1.8)$, $(0.5, -1.6)$, $(1, -2.3)$, $(1.5, 0)$, $(2, 1.3)$, $(2.5, 1.9)$, $(3, 3.1)$.

```
data = {{-1, -2.7}, {-0.5, -2.5}, {0, -1.8}, {0.5, -1.6},      (Mathematica)
       {1, -2.3}, {1.5, 0}, {2, 1.3}, {2.5, 1.9}, {3, 3.1}}
grdata = ListPlot[data]
```

The most commonly used options for **ListPlot** cause the dots to be enlarged or the dots to be connected with straight line segments. Although these two commands cannot be combined in a single call to **PlotStyle**, the same effect can be achieved as shown below.

gr0 = ListPlot[data, PlotStyle–>PointSize[0.02]]   (*Mathematica*)

gr1 = ListPlot[data, PlotJoined–>True]

Show[gr0, gr1]

In **Maple** a scatter plot is obtained using the `plots` package.

```
data:=[[-1,-2.7],[-0.5,-2.5],[0,-1.8],      (Maple)
    [0.5,-1.6],[1,-2.3],[1.5,0],[2,1.3],
    [2.5,1.9],[3,3.1]];
with(plots);
pointplot(data);
```

In **Maple** it is possible to show the points in a variety of symbols and symbol sizes. You can also connect the points with straight line segments. You cannot, however, do both in a single plot. The routine below shows how to accomplish this. It is assumed that the input data is as above and that the `plots` package has been activated.

```
gr0:=pointplot(data,style=point,      (Maple)
    symbol=circle);
gr1:=pointplot(data,style=line);
display([gr0,gr1]);
```

Algorithms that fit relatively simple functions to a set of data points are not difficult to develop but they tend to involve a large number of routine computations. This means that a CAS is particularly well suited to implementing a curve-fitting algorithm. The following routines use a least squares method to fit a line to the data set given above over the interval $[-1, 3]$. The resulting line is then plotted over the data points.

linfit = fit[data, {1, x}, x]   (*Mathematica*)

gr2 = Plot[linfit, {x, -1, 3}]

Show[gr0, gr2]

The format of **data** in the routines above is as a set of points with two coordinates. In **Maple**, to apply a least squares fit of a line to the set of points, we need to reformat the data as a set of

x-coordinates, which we will call **xdata**, and a set of y-coordinates, which we will call **ydata**. To accomplish this we first use the **nops** command to count the number of data points. Then we activate the **stats** pacakage which contains commands for fitting a line to a set of points in the plane.

```
n:=nops(data);                                          (Maple)
xdata:=[seq(data[i,1],i=1..n)];
ydata:=[seq(data[i,2],i=1..n)];
with(stats);
linfit:=fit[leastsquare[[x,y],y=a*x+b,{a,b}]]([xdata,ydata]);
gr3:=implicitplot(linfit,x=-1..3,y=-5..5);
display([gr0,gr3]);
```

In the above *Maple* routine, **implicitplot**, which is part of the **plots** package, is used because the result of the **fit** command is an equation rather than an expression or function. The range of the equation, in this case **y=-5..5**, is specified to be large enough to include all of the data points.

Exercises 3.2   *Hints, Suggestions, Solutions, and Examples*

3. From $dP/dt = P\left(10^{-1} - 10^{-7}P\right)$ and $P(0) = 5000$ we obtain $P = 500/(0.0005 + 0.0995e^{-0.1t})$ so that $P \to 1{,}000{,}000$ as $t \to \infty$. If $P(t) = 500{,}000$ then $t = 52.9\,$months.

5. (*Suggestion*) Notice that no initial population P_0 is specified in the questions on whether the fish population becomes extinct in finite time $t = T$. If there exists such a time, determine which interval

$$0 < P_0 < 1, \quad 1 < P_0 < 4, \quad \text{and} \quad P_0 > 4$$

leads to extinction and then express T in terms of P_0.

6. Solving $P(5 - P) - \frac{25}{4} = 0$ for P we obtain the equilibrium solution $P = \frac{5}{2}$. For $P \neq \frac{5}{2}$, $dP/dt < 0$. Thus, if $P_0 < \frac{5}{2}$, the population becomes extinct (otherwise there would be another equilibrium solution.) Using separation of variables to solve the IVP, we get

$$P(t) = [4P_0 + (10P_0 - 25)t]/[4 + (4P_0 - 10)t].$$

To find when the population becomes extinct for $P_0 < \frac{5}{2}$ we solve $P(t) = 0$ for t. We see that the time of extinction is $t = 4P_0/5(5 - 2P_0)$.

7. (*Suggestion*) Solve the DE by separation of variables. Use completion of the square to carry out the integration:

$$\int \frac{1}{P^2 - 5P + 7} \, dP.$$

8. (*Suggestion*) Solve the DE by separation of variables. The integral $\int dP/P(a - b\ln P)$ may not look familiar, but after rewriting as

$$\int \frac{dP/P}{a - b\ln P},$$

we see that it is of the form $\int du/u$ where $u = a - b\ln P$.

9. Let $X = X(t)$ be the amount of C at time t and $dX/dt = k(120 - 2X)(150 - X)$. If $X(0) = 0$ and $X(5) = 10$, then

$$X(t) = \frac{150 - 150e^{180kt}}{1 - 2.5e^{180kt}},$$

where $k = .0001259$ and $X(20) = 29.3$ grams. Now by L'Hôpital's rule, $X \to 60$ as $t \to \infty$, so that the amount of $A \to 0$ and the amount of $B \to 30$ as $t \to \infty$.

10-11. (*Hint*) After separating variables, integration will involve the form

$$\int u^n \, du = \frac{u^{n+1}}{n+1} + c, \quad n \neq -1.$$

12. We can use separation of variables to solve the differential equation. This gives

$$\frac{dh}{\sqrt{h}} = -c \frac{A_h}{A_w} \sqrt{2g} \, dt$$

$$2\sqrt{h} = -\frac{0.6A_h}{A_w} \sqrt{64} \, t + C \qquad (c = 0.6, g = 32).$$

When $t = 0$, we have $h = 10$ and $A_w = \pi r^2 = 4\pi$, so $C = 2\sqrt{10}$, and

$$h = \left(\sqrt{10} - \frac{2.4A_h}{A_w} t \right)^2.$$

Now, $A_w = \pi r^2 = 4\pi$, and the radius of the hole is $\frac{1}{2}$ in. or $\frac{1}{24}$ ft. so $A_h = \pi/576$. Setting $h = 0$ and solving for t, we find

$$t = \frac{A_w \sqrt{10}}{2.4A_h} = \frac{4\pi\sqrt{10}}{2.4\pi/576} = 960\sqrt{10} \approx 3035.79s.$$

Dividing by 60 to get the number of minutes it takes for the tank to empty, we find $t = 50.6$ min.

62

15. (a) After separating variables we obtain

$$\frac{m\,dv}{mg - kv^2} = dt$$

$$\frac{1}{g}\frac{dv}{1 - (\sqrt{k}\,v/\sqrt{mg}\,)^2} = dt$$

$$\frac{\sqrt{mg}}{\sqrt{k}\,g}\frac{\sqrt{k/mg}\,dv}{1 - (\sqrt{k}\,v/\sqrt{mg}\,)^2} = dt$$

$$\sqrt{\frac{m}{kg}}\,\tanh^{-1}\frac{\sqrt{k}\,v}{\sqrt{mg}} = t + c$$

$$\tanh^{-1}\frac{\sqrt{k}\,v}{\sqrt{mg}} = \sqrt{\frac{kg}{m}}\,t + c_1.$$

Thus the velocity at time t is

$$v(t) = \sqrt{\frac{mg}{k}}\,\tanh\left(\sqrt{\frac{kg}{m}}\,t + c_1\right).$$

Setting $t = 0$ and $v = v_0$ we find $c_1 = \tanh^{-1}(\sqrt{k}\,v_0/\sqrt{mg}\,)$.

(b) Since $\tanh t \to 1$ as $t \to \infty$, we have $v \to \sqrt{mg/k}$ as $t \to \infty$.

(c) Integrating the expression for $v(t)$ in part (a) we obtain an integral of the form $\int du/u$:

$$s(t) = \sqrt{\frac{mg}{k}}\int \tanh\left(\sqrt{\frac{kg}{m}}\,t + c_1\right)dt = \frac{m}{k}\ln\left[\cosh\left(\sqrt{\frac{kg}{m}}\,t + c_1\right)\right] + c_2.$$

Setting $t = 0$ and $s = 0$ we find $c_2 = -(m/k)\ln(\cosh c_1)$, where c_1 is given in part (a).

17. (*Hint*) In this manual, see the discussion on Archimedes' principle in Problem 18 of Exercises 1.3. Also, when separating variables, write

$$\frac{m\,dv}{(mg - \rho V) - (\sqrt{k}\,v)^2} = dt.$$

18. (a) Writing the equation in the form $(x - \sqrt{x^2 + y^2}\,)dx + y\,dy = 0$ we identify $M = x - \sqrt{x^2 + y^2}$ and $N = y$. Since M and N are both homogeneous functions of degree 1 we use the substitution

$y = ux$. It follows that

$$\left(x - \sqrt{x^2 + u^2 x^2}\right) dx + ux(u\, dx + x\, du) = 0$$

$$x\left[1 - \sqrt{1 + u^2} + u^2\right] dx + x^2 u\, du = 0$$

$$-\frac{u\, du}{1 + u^2 - \sqrt{1 + u^2}} = \frac{dx}{x}$$

$$\frac{u\, du}{\sqrt{1 + u^2}\,(1 - \sqrt{1 + u^2})} = \frac{dx}{x}.$$

Letting $w = 1 - \sqrt{1 + u^2}$ we have $dw = -u\, du/\sqrt{1 + u^2}$ so that

$$-\ln\left|1 - \sqrt{1 + u^2}\right| = \ln|x| + c$$

$$\frac{1}{1 - \sqrt{1 + u^2}} = c_1 x$$

$$1 - \sqrt{1 + u^2} = -\frac{c_2}{x} \qquad (-c_2 = 1/c_1)$$

$$1 + \frac{c_2}{x} = \sqrt{1 + \frac{y^2}{x^2}}$$

$$1 + \frac{2c_2}{x} + \frac{c_2^2}{x^2} = 1 + \frac{y^2}{x^2}.$$

Solving for y^2 we have

$$y^2 = 2c_2 x + c_2^2 = 4\left(\frac{c_2}{2}\right)\left(x + \frac{c_2}{2}\right)$$

which is a family of parabolas symmetric with respect to the x-axis with vertex at $(-c_2/2, 0)$ and focus at the origin.

(b) Let $u = x^2 + y^2$ so that

$$\frac{du}{dx} = 2x + 2y\frac{dy}{dx}.$$

Then

$$y\frac{dy}{dx} = \frac{1}{2}\frac{du}{dx} - x$$

and the DE can be written in the form

$$\frac{1}{2}\frac{du}{dx} - x = -x + \sqrt{u} \quad \text{or} \quad \frac{1}{2}\frac{du}{dx} = \sqrt{u}.$$

64

Separating variables and integrating gives

$$\frac{du}{2\sqrt{u}} = dx$$

$$\sqrt{u} = x + c$$

$$u = x^2 + 2cx + c^2$$

$$x^2 + y^2 = x^2 + 2cx + c^2$$

$$y^2 = 2cx + c^2.$$

20d. (*Hint*) Examine $\lim_{t\to\infty} h(t)$ and remember how to rationalize a numerator.

21. (*Suggestion*) Read **Use of Computers** in this section for information on how to use a CAS to draw scatter plots and fit a line to data points.

25. (*Hint*) Here, you should use use

$$\int \frac{du}{a^2 - u^2} = \frac{1}{2a} \ln\left|\frac{u+a}{u-a}\right| + c$$

when evaluating $\int dv/(mg - v^2)$. Also, do not be discouraged by messy algebra.

Section 3.3

Modeling with Systems of First-Order DEs

The terminology and concepts listed below provide an outline of the main ideas encountered in this section. These can be useful when preparing for a quiz or test.

Terminology and Concepts

- systems of first-order DEs – linear and nonlinear

- radioactive decay series

- mixture models involving a system of DEs

- Lotka-Volterra predator-prey model

- competition models

- electrical networks involving a system of first-order DEs

The basic skills listed below summarize the more mechanical types of problems encountered in the exercise set for this section.

Basic Skills

- solve a system of DEs where one of the equations involves only a single unknown
- construct a mathematical model for mixtures involving multiple tanks
- use a computer program to numerically analyze solutions of a system of DEs

Use of Computers
There are a number of computer programs that can solve a DE or system of DEs numerically and graph the resulting numerical solution curves. We show below how *Mathematica* and *Maple* can produce graphs of numerical solutions.

In *Mathematica* **NDSolve** is used to obtain a numerical solution of an IVP which can then be graphed.

```
Clear[y]                                                    (Mathematica)
sol=Flatten[NDSolve[{y'[x]==Exp[x/2]Sin[x^2], y[0]==0}, y, {x, -5, 5}]];
y[x_]:=Evaluate[y[x]/.sol]
y[x]
Plot[y[x], {x, -5, 5}]
```

Note that the interval $[-5, 5]$ specified in the **Plot** command can be no larger than the interval used in the **NDSolve** command. Following is an example of how *Mathematica* can be used to obtain a numerical solution of a system of DEs with initial conditions.

```
Clear[y]
sol=Flatten[NDSolve[{x'[t]== -0.3x[t] + 0.1x[t]y[t],
y'[t]==0.2y[t] - 0.1x[t]y[t], x[0]==6, y[0]==4}, {x, y}, {t, -2, 2}]];
x[t_]:=Evaluate[x[t]/.sol]
y[t_]:=Evaluate[y[t]/.sol]
x[t]
y[t]
Plot[{x[t], y[t]}, {t, -2, 2}]
```

In *Maple* there are two ways to plot the numerical solution curve of an IVP. The first way shown below returns a procedure that can be used with the **plots** package to graph the solution. It is also possible to define a function that can evaluate the solution at any point.

```
sol:=dsolve({diff(y(x),x)=exp(x/t)*sin(x^2),    (Maple)
y(0)=0},{y(x)}, type=numeric);
with(plots);
odeplots(sol,[x,y(x)],-5..5);
y1:=x1->eval(y(x),sol(x1));
y1(2.67);
```

The second way to plot a numerical solution of the DE uses the **DEtools** package.

```
with(DEtools);                                            (Maple)
DEplot(diff(y(x),x)=exp(x/2)*sin(x^2),y(x),-5..5,
    [[y(0)=0]],arrows=NONE,stepsize=0.05);
```

The **arrows** option suppresses a direction field in the background of the graph and the **stepsize** option smooths the graph out.

Maple can also graph the numerical solutions of a system of DEs with initial conditions. In this case we first specify the DEs and initial conditions before calling the **DEtools** package.

```
eq1:=diff(x(t),t)=-0.3*x(t)+0.1*x(t)*y(t);       (Maple)
eq2:=diff(y(t),t)=0.2*y(t)-0.1*x(t)*y(t);
with(DEtools);
gr1:=DEplot({eq1,eq2},[x(t),y(t)],-5..5,[[x(0)=6,y(0)=4]],
    scene=[t,x]):
gr2:=DEplot({eq1,eq2},[x(t),y(t)],-5..5,[[x(0)=6,y(0)=4]],
    scene=[t,y]):
display([gr1,gr2]);
```

Note the use of colons, rather than the usual **semicolons**, in the lines above defining **gr1** and **gr2**. This prevents the display of a very long list of uninteresting expressions.

Exercises 3.3   *Hints, Suggestions, Solutions, and Examples*

3. The amounts x and y are the same at about $t = 5$ days. The amounts x and z are the same at about $t = 20$ days. The amounts y and z are the same at about $t = 147$ days. The time when y and z are the same makes sense because most of A and half of B are gone, so half of C should have been formed.

6. Let x_1, x_2, and x_3 be the amounts of salt in tanks A, B, and C, respectively, so that

$$x_1' = \frac{1}{100}x_2 \cdot 2 - \frac{1}{100}x_1 \cdot 6 = \frac{1}{50}x_2 - \frac{3}{50}x_1$$

$$x_2' = \frac{1}{100}x_1 \cdot 6 + \frac{1}{100}x_3 - \frac{1}{100}x_2 \cdot 2 - \frac{1}{100}x_2 \cdot 5 = \frac{3}{50}x_1 - \frac{7}{100}x_2 + \frac{1}{100}x_3$$

$$x_3' = \frac{1}{100}x_2 \cdot 5 - \frac{1}{100}x_3 - \frac{1}{100}x_3 \cdot 4 = \frac{1}{20}x_2 - \frac{1}{20}x_3.$$

9. Zooming in on the graph it can be seen that the populations are first equal at about $t = 5.6$. The approximate periods of x and y are both 45.

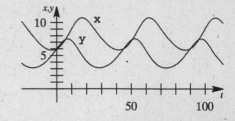

10-11. (*Suggestion*) See **Use of Computers** in this section for *Mathematica* and *Maple* routines that can be used to graph numerical solutions of a system of DEs.

12. By Kirchhoff's first law we have $i_1 = i_2 + i_3$. By Kirchhoff's second law, on each loop we have $E(t) = Li_1' + R_1i_2$ and $E(t) = Li_1' + R_2i_3 + q/C$ so that $q = CR_1i_2 - CR_2i_3$. Then $i_3 = q' = CR_1i_2' - CR_2i_3$ so that the system is

$$Li_2' + Li_3' + R_1i_2 = E(t)$$

$$-R_1i_2' + R_2i_3' + \frac{1}{C}i_3 = 0.$$

15. We first note that $s(t) + i(t) + r(t) = n$. Now the rate of change of the number of susceptible persons, $s(t)$, is proportional to the number of contacts between the number of people infected and the number who are susceptible; that is, $ds/dt = -k_1si$. We use $-k_1 < 0$ because $s(t)$ is decreasing. Next, the rate of change of the number of persons who have recovered is proportional to the number infected; that is, $dr/dt = k_2i$ where $k_2 > 0$ since r is increasing. Finally, to obtain di/dt we use

$$\frac{d}{dt}(s + i + r) = \frac{d}{dt}n = 0.$$

This gives

$$\frac{di}{dt} = -\frac{dr}{dt} - \frac{ds}{dt} = -k_2i + k_1si.$$

The system of differential equations is then

$$\frac{ds}{dt} = -k_1si$$

$$\frac{di}{dt} = -k_2i + k_1si$$

$$\frac{dr}{dt} = k_2i.$$

A reasonable set of initial conditions is $i(0) = i_0$, the number of infected people at time 0, $s(0) = n - i_0$, and $r(0) = 0$.

16. (*Suggestion*) See **Use of Computers** in this section for *Mathematica* and *Maple* routines that can be used to graph numerical solutions of a system of DEs.

18. (*Hint*) Let $z(t) = x(t) - y(t)$ and solve the resulting equation.

19. (*Suggestion*) See **Use of Computers** in this section for *Mathematica* and *Maple* routines that can be used to graph numerical solutions of a system of DEs.

Chapter 3 in Review

Hints, Suggestions, Solutions, and Examples

3. From $\dfrac{dP}{dt} = 0.018P$ and $P(0) = 4$ billion we obtain $P = 4e^{0.018t}$ so that $P(45) = 8.99$ billion.

6. From $V\, dC/dt = kA(C_s - C)$ and $C(0) = C_0$ we obtain $C = C_s + (C_0 - C_s)e^{-kAt/V}$.

8. (*Hint*) In part (a) use separation of variables, and in part (b) write the DE as

$$\frac{dT}{dt} = [(T_m + (T - T_m))^4 - T_m^4] = kT_m^4 \left[\left(1 + \frac{T - T_m}{T_m}\right)^4 - 1\right].$$

9. We first solve $(1 - t/10)di/dt + 0.2i = 4$. Separating variables we obtain $di/(40 - 2i) = dt/(10 - t)$. Then

$$-\frac{1}{2}\ln|40 - 2i| = -\ln|10 - t| + c \quad \text{or} \quad \sqrt{40 - 2i} = c_1(10 - t).$$

Since $i(0) = 0$ we must have $c_1 = 2/\sqrt{10}$. Solving for i we get $i(t) = 4t - \frac{1}{5}t^2$, $0 \le t < 10$. For $t \ge 10$ the equation for the current becomes $0.2i = 4$ or $i = 20$. Thus

$$i(t) = \begin{cases} 4t - \frac{1}{5}t^2, & 0 \le t < 10 \\ 20, & t \ge 10. \end{cases}$$

The graph of $i(t)$ is given in the figure.

12. In tank A the salt input is

$$\left(7\,\frac{\text{gal}}{\text{min}}\right)\left(2\,\frac{\text{lb}}{\text{gal}}\right) + \left(1\,\frac{\text{gal}}{\text{min}}\right)\left(\frac{x_2}{100}\,\frac{\text{lb}}{\text{gal}}\right) = \left(14 + \frac{1}{100}x_2\right)\frac{\text{lb}}{\text{min}}.$$

The salt output is

$$\left(3\,\frac{\text{gal}}{\text{min}}\right)\left(\frac{x_1}{100}\,\frac{\text{lb}}{\text{gal}}\right) + \left(5\,\frac{\text{gal}}{\text{min}}\right)\left(\frac{x_1}{100}\,\frac{\text{lb}}{\text{gal}}\right) = \frac{2}{25}x_1\,\frac{\text{lb}}{\text{min}}.$$

In tank B the salt input is

$$\left(5\,\frac{\text{gal}}{\text{min}}\right)\left(\frac{x_1}{100}\,\frac{\text{lb}}{\text{gal}}\right) = \frac{1}{20}\,x_1\,\frac{\text{lb}}{\text{min}}\,.$$

The salt output is

$$\left(1\,\frac{\text{gal}}{\text{min}}\right)\left(\frac{x_2}{100}\,\frac{\text{lb}}{\text{gal}}\right) + \left(4\,\frac{\text{gal}}{\text{min}}\right)\left(\frac{x_2}{100}\,\frac{\text{lb}}{\text{gal}}\right) = \frac{1}{20}\,x_2\,\frac{\text{lb}}{\text{min}}\,.$$

The system of differential equations is then

$$\frac{dx_1}{dt} = 14 + \frac{1}{100}x_2 - \frac{2}{25}x_1$$

$$\frac{dx_2}{dt} = \frac{1}{20}x_1 - \frac{1}{20}x_2.$$

4 Higher-Order Differential Equations

Section 4.1
Preliminary Theory—Linear Equations

The terminology and concepts listed below provide an outline of the main ideas encountered in this section. These can be useful when preparing for a quiz or test.

Terminology and Concepts

- linear nth-order initial-value problem
- existence-uniqueness theorem for linear nth-order IVPs
- boundary-value problems for linear nth-order DEs
- no existence-uniqueness theorem for linear nth-order boundary-value problems
- homogeneous linear nth-order DEs
- representation of a DE using differential operators
- superposition principles
- linear dependence and linear independence of a set of functions
- Wronskian
- fundamental set of solutions
- general solution of a linear nth-order DE
- particular solution of a nonhomogeneous linear nth-order DE
- complementary function for a nonhomogeneous linear nth-order DE

The basic skills listed below summarize the more mechanical types of problems encountered in the exercise set for this section.

Basic Skills

- given the general solution of a linear nth-order DE, find a solution satisfying n initial conditions

- given a two-parameter family of solutions of a linear second-order DE, determine a member or members of the family satisfying specified boundary conditions (or show that no solution exists)

- determine whether a given set of functions is linearly independent on an interval

- show that a given set of functions forms a fundamental set of solutions on an interval

Exercises 4.1   *Hints, Suggestions, Solutions, and Examples*

3. From $y = c_1 x + c_2 x \ln x$ we find $y' = c_1 + c_2(1 + \ln x)$. Then $y(1) = c_1 = 3$, $y'(1) = c_1 + c_2 = -1$ so that $c_1 = 3$ and $c_2 = -4$. The solution is $y = 3x - 4x \ln x$.

6. In this case we have $y(0) = c_1 = 0$, $y'(0) = 2c_2 \cdot 0 = 0$ so $c_1 = 0$ and c_2 is arbitrary. Two solutions are $y = x^2$ and $y = 2x^2$.

9. Since $a_2(x) = x - 2$ and $x_0 = 0$ the problem has a unique solution for $-\infty < x < 2$.

12. In this case we have $y(0) = c_1 = 1$, $y'(1) = 2c_2 = 6$ so that $c_1 = 1$ and $c_2 = 3$. The solution is $y = 1 + 3x^2$.

15. Since $(-4)x + (3)x^2 + (1)(4x - 3x^2) = 0$ the set of functions is linearly dependent.

18. Since $(1)\cos 2x + (1)1 + (-2)\cos^2 x = 0$ the set of functions is linearly dependent.

21. Suppose $c_1(1 + x) + c_2 x + c_3 x^2 = 0$. Then $c_1 + (c_1 + c_2)x + c_3 x^2 = 0$ and so $c_1 = 0$, $c_1 + c_2 = 0$, and $c_3 = 0$. Since $c_1 = 0$ we also have $c_2 = 0$. Thus, the set of functions is linearly independent.

23-30. (*Hint*) Remember that in order for a set of n functions to be a fundamental set of solutions of a linear nth-order DE they must *both* satisfy the DE and be linearly independent. Furthermore, if they satisfy the DE, then linear independence can be verified using the Wronskian.

24. The functions satisfy the differential equation and are linearly independent since

$$W(\cosh 2x, \sinh 2x) = 2$$

for $-\infty < x < \infty$. The general solution is

$$y = c_1 \cosh 2x + c_2 \sinh 2x.$$

27. The functions satisfy the differential equation and are linearly independent since

$$W\left(x^3, x^4\right) = x^6 \neq 0$$

for $0 < x < \infty$. The general solution on this interval is

$$y = c_1 x^3 + c_2 x^4.$$

30. The functions satisfy the differential equation and are linearly independent since

$$W(1, x, \cos x, \sin x) = 1$$

for $-\infty < x < \infty$. The general solution on this interval is

$$y = c_1 + c_2 x + c_3 \cos x + c_4 \sin x.$$

33. The functions $y_1 = e^{2x}$ and $y_2 = xe^{2x}$ form a fundamental set of solutions of the associated homogeneous equation, and $y_p = x^2 e^{2x} + x - 2$ is a particular solution of the nonhomogeneous equation.

36. (a) $y_{p_1} = 5$

(b) $y_{p_2} = -2x$

(c) $y_p = y_{p_1} + y_{p_2} = 5 - 2x$

(d) $y_p = \frac{1}{2} y_{p_1} - 2 y_{p_2} = \frac{5}{2} + 4x$

Section 4.2

Reduction of Order

The terminology and concepts listed below provide an outline of the main ideas encountered in this section. These can be useful when preparing for a quiz or test.

Terminology and Concepts

- reduction of order

The basic skills listed below summarize the more mechanical types of problems encountered in the exercise set for this section.

Basic Skills

- given a solution y_1 of a linear second-order homogeneous DE, find a second solution y_2 using reduction of order
- given a solution y_1 of a linear second-order homogeneous DE, find a second solution y_2 using formula (5) in the text

Reduction of order is used below in Problems 3 and 6 to find a second solution.

3. Define $y = u(x)\cos 4x$ so

$$y' = -4u\sin 4x + u'\cos 4x, \quad y'' = u''\cos 4x - 8u'\sin 4x - 16u\cos 4x$$

and

$$y'' + 16y = (\cos 4x)u'' - 8(\sin 4x)u' = 0 \quad \text{or} \quad u'' - 8(\tan 4x)u' = 0.$$

If $w = u'$ we obtain the linear first-order equation $w' - 8(\tan 4x)w = 0$ which has the integrating factor $e^{-8\int \tan 4x\, dx} = \cos^2 4x$. Now

$$\frac{d}{dx}[(\cos^2 4x)w] = 0 \quad \text{gives} \quad (\cos^2 4x)w = c.$$

Therefore $w = u' = c\sec^2 4x$ and $u = c_1\tan 4x$. A second solution is $y_2 = \tan 4x\cos 4x = \sin 4x$.

6. Define $y = u(x)e^{5x}$ so

$$y' = 5e^{5x}u + e^{5x}u', \quad y'' = e^{5x}u'' + 10e^{5x}u' + 25e^{5x}u$$

and

$$y'' - 25y = e^{5x}(u'' + 10u') = 0 \quad \text{or} \quad u'' + 10u' = 0.$$

If $w = u'$ we obtain the linear first-order equation $w' + 10w = 0$ which has the integrating factor $e^{10\int dx} = e^{10x}$. Now

$$\frac{d}{dx}[e^{10x}w] = 0 \quad \text{gives} \quad e^{10x}w = c.$$

Therefore $w = u' = ce^{-10x}$ and $u = c_1 e^{-10x}$. A second solution is $y_2 = e^{-10x}e^{5x} = e^{-5x}$.

Formula (5) from the text is used in Problems 9, 12, and 15.

9. Identifying $P(x) = -7/x$ we have

$$y_2 = x^4 \int \frac{e^{-\int(-7/x)\,dx}}{x^8}\, dx = x^4 \int \frac{1}{x}\, dx = x^4 \ln|x|.$$

A second solution is $y_2 = x^4 \ln|x|$.

12. Identifying $P(x) = 0$ we have

$$y_2 = x^{1/2}\ln x \int \frac{e^{-\int 0\,dx}}{x(\ln x)^2}\, dx = x^{1/2}\ln x\left(-\frac{1}{\ln x}\right) = -x^{1/2}.$$

A second solution is $y_2 = x^{1/2}$.

15. Identifying $P(x) = 2(1+x)/\left(1 - 2x - x^2\right)$ we have

$$y_2 = (x+1) \int \frac{e^{-\int 2(1+x)dx/\left(1-2x-x^2\right)}}{(x+1)^2} \, dx = (x+1) \int \frac{e^{\ln\left(1-2x-x^2\right)}}{(x+1)^2} \, dx$$

$$= (x+1) \int \frac{1 - 2x - x^2}{(x+1)^2} \, dx = (x+1) \int \left[\frac{2}{(x+1)^2} - 1\right] dx$$

$$= (x+1) \left[-\frac{2}{x+1} - x\right] = -2 - x^2 - x.$$

A second solution is $y_2 = x^2 + x + 2$.

17. (*Hint*) Look for a particular solution having the form $y_p = A$.

18. Define $y = u(x) \cdot 1$ so

$$y' = u', \quad y'' = u'' \quad \text{and} \quad y'' + y' = u'' + u' = 1.$$

If $w = u'$ we obtain the linear first-order equation $w' + w = 1$ which has the integrating factor $e^{\int dx} = e^x$. Now

$$\frac{d}{dx}[e^x w] = e^x \quad \text{gives} \quad e^x w = e^x + c.$$

Therefore $w = u' = 1 + ce^{-x}$ and $u = x + c_1 e^{-x} + c_2$. The general solution is

$$y = u = x + c_1 e^{-x} + c_2.$$

19. (*Hint*) Look for a particular solution having the form $y_p = Ae^{3x}$.

20. (*Hint*) Look for a particular solution having the form $y_p = Ax + B$.

22. (*Hint*) Recall the infinite series representation of the exponential function: $e^x = \sum_{n=0}^{\infty}(1/n!)x^n$.

Section 4.3
Homogeneous Linear Equations with Constant Coefficients

The terminology and concepts listed below provide an outline of the main ideas encountered in this section. These can be useful when preparing for a quiz or test.

Terminology and Concepts

- auxiliary equation

The basic skills listed below summarize the more mechanical types of problems encountered in the exercise set for this section.

- find the general solution of a homogeneous linear DE with constant coefficients when the roots of the auxiliary equation are real and distinct, real and repeated, conjugate complex, and repeated conjugate complex

- find the solution of an nth-order homogeneous linear DE with constant coefficients and n initial conditions

Factorization of Polynomials   In Section 4.3 in the text the solution method for homogeneous linear DEs with constant coefficents relies on your ability to solve polynomial equations. For example, you know that a quadratic polynomial equation

$$am^2 + bm + c = 0, \quad a \neq 0, \ a, b, c \text{ real constants,}$$

can always be solved using the quadratic formula. The two roots m_1 and m_2 are

$$m_1 = \frac{-b + \sqrt{b^2 - 4ac}}{2a} \quad \text{and} \quad m_2 = \frac{-b - \sqrt{b^2 - 4ac}}{2a}.$$

The roots m_1 and m_2 are also called the **zeros** of the polynomial $am^2 + bm + c$. Note that m_1 and m_2 are real and distinct ($m_1 \neq m_2$) if $b^2 - 4ac > 0$, real and equal ($m_1 = m_2$) if $b^2 - 4ac = 0$, and finally, m_1 and m_2 are complex conjugates ($m_1 = \alpha + i\beta$, $m_1 = \alpha - i\beta$, α and β are positive real numbers) if $b^2 - 4ac < 0$. The Factor Theorem of Algebra states that finding zeros of a polynomial and factoring the polynomial are equivalent problems. Thus, in the case of a quadratic polynomial $am^2 + bm + c$ we can write

$$am^2 + bm + c = a(m - m_1)(m - m_2).$$

A polynomial of degree n, $a_n m^n + a_{n-1} m^{n-1} + \cdots + a_1 m + a_0$ with real coefficients has n zeros (counting multiplicities) and hence can be factored as

$$a_n m^n + a_{n-1} m^{n-1} + \cdots + a_1 m + a_0 = a_n(m - m_1)(m - m_2) \cdots (m - m_n).$$

Finding the zeros requires that we find roots of a polynomial equation of degree > 2, and *that* is the fundamental problem here. Analogous to quadratic equations, cubic and quartic polynomial equations can be solved by (complicated) formulas involving radicals, but there exist no such formulas for solving polynomial equations of degree $n > 4$. But if the polynomial has *integer* coefficients and *if p/q is a rational zero (p and q are integers and $q \neq 0$), then it can be proved that

$$p \text{ must be an integer factor of } a_0,$$

and

$$q \text{ must be an integer factor of } a_n.$$

By forming all possible quotients of each factor of a_0 to each factor of a_n, we can construct a list of all *possible* rational zeros. For example, consider the fourth-degree polynomial equation $3m^4 - 10m^3 - 3m^2 + 8m - 2 = 0$. We identify $a_0 = -2$ and $a_n = 3$ and list all integer factors of a_0 and a_n, respectively:

$$p : \pm 1, \pm 2 \qquad \text{and} \qquad q : \pm 1, \pm 3.$$

The eight possible rational roots of the equation are then

$$\frac{p}{q} : -1, 1, -\frac{1}{3}, \frac{1}{3}, -2, 2, -\frac{2}{3}, \frac{2}{3}.$$

To determine which, if any, of these numbers are roots, we could use brute force substitution of a number p/q into the polynomial to see whether it satisfies the equation. We could also use long division. If the term $m - p/q$ divides the polynomial evenly, that is, the remainder r is zero, then $m - p/q$ is a factor and hence p/q is a root. A short-hand version of long division, called **synthetic division**, is an efficient way of proceeding. We begin by testing -1 by synthetically dividing by $m - (-1) = m + 1$:

$$
\begin{array}{r|rrrrr}
-1 & 3 & -10 & -3 & 8 & -2 \\
   &   & -3 & 13 & -10 & 2 \\
\hline
   & 3 & -13 & 10 & -2 & \boxed{0 = r}
\end{array}
$$

Since the remainder upon division is $r = 0$, we can conclude $m + 1$ is a factor of the polynomial and so -1 is a root. Now the division shows that

$$3m^4 - 10m^3 - 3m^2 + 8m - 2 = (m+1)(3m^3 - 13m^2 + 10m - 2).$$

Observe that if the original polynomial has additional rational zeros they must be zeros of the second, or cubic, factor $3m^3 - 13m^2 + 10m - 2$. Thus we continue the testing by dividing this factor by $m - 1$:

$$
\begin{array}{r|rrrr}
1 & 3 & -13 & 10 & -2 \\
  &   & 3   & -10 & 0 \\
\hline
  & 3 & -10 & 0 & \boxed{-2 = r}
\end{array}
$$

The foregoing division shows that $m - 1$ is not a factor since the remainder r is not zero, and so 1 is not a rational root. Checking $\frac{1}{3}$ we have

$$\frac{1}{3} \begin{array}{|cccc} 3 & -13 & 10 & -2 \\ & 1 & -4 & 2 \\ \hline 3 & -12 & 6 & \boxed{0} = r \end{array}$$

Thus, $\frac{1}{3}$ is a root, and we now have the factorization

$$3m^4 - 10m^3 - 3m^2 + 8m - 2 = (m+1)\left(m - \frac{1}{3}\right)(3m^2 - 12m + 6)$$

$$= (m+1)\left(m - \frac{1}{3}\right)(3)(m^2 - 4m + 2)$$

$$= (m+1)(3m - 1)(m^2 - 4m + 2).$$

Since the factor $m^2 - 4m + 2$ discovered by the division is quadratic, the remaining roots, $2 + \sqrt{2}$ and $2 - \sqrt{2}$, can be determined from the quadratic formula. Therefore, the original polynomial equation has four real roots: two rational and two irrational.

Exercises 4.3   *Hints, Suggestions, Solutions, and Examples*

3. From $m^2 - m - 6 = 0$ we obtain $m = 3$ and $m = -2$ so that $y = c_1 e^{3x} + c_2 e^{-2x}$.

6. From $m^2 - 10m + 25 = 0$ we obtain $m = 5$ and $m = 5$ so that $y = c_1 e^{5x} + c_2 x e^{5x}$.

9. From $m^2 + 9 = 0$ we obtain $m = 3i$ and $m = -3i$ so that $y = c_1 \cos 3x + c_2 \sin 3x$.

12. From $2m^2 + 2m + 1 = 0$ we obtain $m = -1/2 \pm i/2$ so that

$$y = e^{-x/2}[c_1 \cos(x/2) + c_2 \sin(x/2)].$$

15. From $m^3 - 4m^2 - 5m = 0$ we obtain $m = 0$, $m = 5$, and $m = -1$ so that

$$y = c_1 + c_2 e^{5x} + c_3 e^{-x}.$$

18. From $m^3 + 3m^2 - 4m - 12 = 0$ we obtain $m = -2$, $m = 2$, and $m = -3$ so that

$$y = c_1 e^{-2x} + c_2 e^{2x} + c_3 e^{-3x}.$$

21. From $m^3 + 3m^2 + 3m + 1 = 0$ we obtain $m = -1$, $m = -1$, and $m = -1$ so that

$$y = c_1 e^{-x} + c_2 x e^{-x} + c_3 x^2 e^{-x}.$$

24. From $m^4 - 2m^2 + 1 = 0$ we obtain $m = 1$, $m = 1$, $m = -1$, and $m = -1$ so that

$$y = c_1 e^x + c_2 x e^x + c_3 e^{-x} + c_4 x e^{-x}.$$

27. From $m^5 + 5m^4 - 2m^3 - 10m^2 + m + 5 = 0$ we obtain $m = -1$, $m = -1$, $m = 1$, and $m = 1$, and $m = -5$ so that

$$u = c_1 e^{-r} + c_2 r e^{-r} + c_3 e^r + c_4 r e^r + c_5 e^{-5r}.$$

30. From $m^2 + 1 = 0$ we obtain $m = \pm i$ so that $y = c_1 \cos\theta + c_2 \sin\theta$. If $y(\pi/3) = 0$ and $y'(\pi/3) = 2$ then

$$\frac{1}{2}c_1 + \frac{\sqrt{3}}{2}c_2 = 0$$

$$-\frac{\sqrt{3}}{2}c_1 + \frac{1}{2}c_2 = 2,$$

so $c_1 = -\sqrt{3}$, $c_2 = 1$, and $y = -\sqrt{3}\cos\theta + \sin\theta$.

33. From $m^2 + m + 2 = 0$ we obtain $m = -1/2 \pm \sqrt{7}\,i/2$ so that $y = e^{-x/2}[c_1 \cos(\sqrt{7}\,x/2) + c_2 \sin(\sqrt{7}\,x/2)]$. If $y(0) = 0$ and $y'(0) = 0$ then $c_1 = 0$ and $c_2 = 0$ so that $y = 0$.

36. From $m^3 + 2m^2 - 5m - 6 = 0$ we obtain $m = -1$, $m = 2$, and $m = -3$ so that

$$y = c_1 e^{-x} + c_2 e^{2x} + c_3 e^{-3x}.$$

If $y(0) = 0$, $y'(0) = 0$, and $y''(0) = 1$ then

$$c_1 + c_2 + c_3 = 0, \quad -c_1 + 2c_2 - 3c_3 = 0, \quad c_1 + 4c_2 + 9c_3 = 1,$$

so $c_1 = -1/6$, $c_2 = 1/15$, $c_3 = 1/10$, and

$$y = -\frac{1}{6}e^{-x} + \frac{1}{15}e^{2x} + \frac{1}{10}e^{-3x}.$$

39. From $m^2 + 1 = 0$ we obtain $m = \pm i$ so that $y = c_1 \cos x + c_2 \sin x$ and $y' = -c_1 \sin x + c_2 \cos x$. From $y'(0) = c_1(0) + c_2(1) = c_2 = 0$ and $y'(\pi/2) = -c_1(1) = 0$ we find $c_1 = c_2 = 0$. A solution of the BVP is $y = 0$.

42. The auxiliary equation is $m^2 - 1 = 0$ which has roots -1 and 1. By (10) in the text the general solution is $y = c_1 e^x + c_2 e^{-x}$. By (11) in the text the general solution is $y = c_1 \cosh x + c_2 \sinh x$. For $y = c_1 e^x + c_2 e^{-x}$ the boundary conditions imply $c_1 + c_2 = 1$, $c_1 e - c_2 e^{-1} = 0$. Solving for c_1 and c_2 we find $c_1 = 1/(1 + e^2)$ and $c_2 = e^2/(1 + e^2)$ so $y = e^x/(1 + e^2) + e^2 e^{-x}/(1 + e^2)$. For $y = c_1 \cosh x + c_2 \sinh x$ the boundary conditions imply $c_1 = 1$, $c_2 = -\tanh 1$, so $y = \cosh x - (\tanh 1)\sinh x$.

45. The auxiliary equation should have a pair of complex roots $\alpha \pm \beta i$ where $\alpha < 0$, so that the solution has the form $e^{\alpha x}(c_1 \cos\beta x + c_2 \sin\beta x)$. Thus, the DE is (e).

48. The DE should have the form $y'' + k^2 y = 0$ where $k = 2$ so that the period of the solution is π. Thus, the DE is (b).

49. (*Hint*) Read the discussion above in this section about the relationship between finding zeros of a polynomial and factoring the polynomial.

<table>
<tr><td>**Section 4.4**</td><td>**Undetermined Coefficients – Superposition Approach**</td></tr>
</table>

The terminology and concepts listed below provide an outline of the main ideas encountered in this section. These can be useful when preparing for a quiz or test.

Terminology and Concepts

- undetermined coefficients
- superposition principle
- multiplication rule

The basic skills listed below summarize the more mechanical types of problems encountered in the exercise set for this section.

Basic Skills

- use undetermined coefficients with the superposition approach to find a particular solution of a nonhomogeneous linear DE with constant coefficients when (1) no function in the assumed particular solution is a solution of the associated homogeneous DE, and (2) a function in the assumed particular solution is also a solution of the associated homogeneous DE

- recognize when the input function (right-hand side of the DE) is appropriate for the method of undetermined coefficients.

Exercises 4.4   *Hints, Suggestions, Solutions, and Examples*

3. From $m^2 - 10m + 25 = 0$ we find $m_1 = m_2 = 5$. Then $y_c = c_1 e^{5x} + c_2 x e^{5x}$ and we assume $y_p = Ax + B$. Substituting into the DE we obtain $25A = 30$ and $-10A + 25B = 3$. Then $A = \frac{6}{5}$, $B = \frac{3}{5}$, $y_p = \frac{6}{5}x + \frac{3}{5}$, and

$$y = c_1 e^{5x} + c_2 x e^{5x} + \frac{6}{5}x + \frac{3}{5}.$$

6. From $m^2 - 8m + 20 = 0$ we find $m_1 = 4 + 2i$ and $m_2 = 4 - 2i$. Then $y_c = e^{4x}(c_1 \cos 2x + c_2 \sin 2x)$ and we assume $y_p = Ax^2 + Bx + C + (Dx + E)e^x$.

Substituting into the DE we obtain

$$2A - 8B + 20C = 0$$

$$-6D + 13E = 0$$

$$-16A + 20B = 0$$

$$13D = -26$$

$$20A = 100.$$

Then $A = 5$, $B = 4$, $C = \frac{11}{10}$, $D = -2$, $E = -\frac{12}{13}$, $y_p = 5x^2 + 4x + \frac{11}{10} + \left(-2x - \frac{12}{13}\right)e^x$ and

$$y = e^{4x}(c_1 \cos 2x + c_2 \sin 2x) + 5x^2 + 4x + \frac{11}{10} + \left(-2x - \frac{12}{13}\right)e^x.$$

9. From $m^2 - m = 0$ we find $m_1 = 1$ and $m_2 = 0$. Then $y_c = c_1 e^x + c_2$ and we assume $y_p = Ax$. Substituting into the DE we obtain $-A = -3$. Then $A = 3$, $y_p = 3x$ and $y = c_1 e^x + c_2 + 3x$.

12. From $m^2 - 16 = 0$ we find $m_1 = 4$ and $m_2 = -4$. Then $y_c = c_1 e^{4x} + c_2 e^{-4x}$ and we assume $y_p = Axe^{4x}$. Substituting into the DE we obtain $8A = 2$. Then $A = \frac{1}{4}$, $y_p = \frac{1}{4}xe^{4x}$ and

$$y = c_1 e^{4x} + c_2 e^{-4x} + \frac{1}{4}xe^{4x}.$$

15. From $m^2 + 1 = 0$ we find $m_1 = i$ and $m_2 = -i$. Then $y_c = c_1 \cos x + c_2 \sin x$ and we assume $y_p = (Ax^2 + Bx)\cos x + (Cx^2 + Dx)\sin x$. Substituting into the DE we obtain $4C = 0$, $2A + 2D = 0$, $-4A = 2$, and $-2B + 2C = 0$. Then $A = -\frac{1}{2}$, $B = 0$, $C = 0$, $D = \frac{1}{2}$, $y_p = -\frac{1}{2}x^2 \cos x + \frac{1}{2}x \sin x$, and

$$y = c_1 \cos x + c_2 \sin x - \frac{1}{2}x^2 \cos x + \frac{1}{2}x \sin x.$$

18. From $m^2 - 2m + 2 = 0$ we find $m_1 = 1 + i$ and $m_2 = 1 - i$. Then $y_c = e^x(c_1 \cos x + c_2 \sin x)$ and we assume $y_p = Ae^{2x}\cos x + Be^{2x}\sin x$. Substituting into the DE we obtain $A + 2B = 1$ and $-2A + B = -3$. Then $A = \frac{7}{5}$, $B = -\frac{1}{5}$, $y_p = \frac{7}{5}e^{2x}\cos x - \frac{1}{5}e^{2x}\sin x$ and

$$y = e^x(c_1 \cos x + c_2 \sin x) + \frac{7}{5}e^{2x}\cos x - \frac{1}{5}e^{2x}\sin x.$$

21. From $m^3 - 6m^2 = 0$ we find $m_1 = m_2 = 0$ and $m_3 = 6$. Then $y_c = c_1 + c_2 x + c_3 e^{6x}$ and we assume $y_p = Ax^2 + B\cos x + C\sin x$. Substituting into the DE we obtain $-12A = 3$, $6B - C = -1$, and $B + 6C = 0$. Then $A = -\frac{1}{4}$, $B = -\frac{6}{37}$, $C = \frac{1}{37}$, $y_p = -\frac{1}{4}x^2 - \frac{6}{37}\cos x + \frac{1}{37}\sin x$, and

$$y = c_1 + c_2 x + c_3 e^{6x} - \frac{1}{4}x^2 - \frac{6}{37}\cos x + \frac{1}{37}\sin x.$$

24. From $m^3 - m^2 - 4m + 4 = 0$ we find $m_1 = 1$, $m_2 = 2$, and $m_3 = -2$. Then $y_c = c_1 e^x + c_2 e^{2x} + c_3 e^{-2x}$ and we assume $y_p = A + Bxe^x + Cxe^{2x}$. Substituting into the DE we obtain $4A = 5$, $-3B = -1$, and $4C = 1$. Then $A = \frac{5}{4}$, $B = \frac{1}{3}$, $C = \frac{1}{4}$, $y_p = \frac{5}{4} + \frac{1}{3}xe^x + \frac{1}{4}xe^{2x}$, and

$$y = c_1 e^x + c_2 e^{2x} + c_3 e^{-2x} + \frac{5}{4} + \frac{1}{3}xe^x + \frac{1}{4}xe^{2x}.$$

27. We have $y_c = c_1 \cos 2x + c_2 \sin 2x$ and we assume $y_p = A$. Substituting into the DE we find $A = -\frac{1}{2}$. Thus $y = c_1 \cos 2x + c_2 \sin 2x - \frac{1}{2}$. From the initial conditions we obtain $c_1 = 0$ and $c_2 = \sqrt{2}$, so $y = \sqrt{2} \sin 2x - \frac{1}{2}$.

30. We have $y_c = c_1 e^{-2x} + c_2 xe^{-2x}$ and we assume $y_p = (Ax^3 + Bx^2)e^{-2x}$. Substituting into the DE we find $A = \frac{1}{6}$ and $B = \frac{3}{2}$. Thus $y = c_1 e^{-2x} + c_2 xe^{-2x} + \left(\frac{1}{6}x^3 + \frac{3}{2}x^2\right)e^{-2x}$. From the initial conditions we obtain $c_1 = 2$ and $c_2 = 9$, so

$$y = 2e^{-2x} + 9xe^{-2x} + \left(\frac{1}{6}x^3 + \frac{3}{2}x^2\right)e^{-2x}.$$

33. We have $x_c = c_1 \cos \omega t + c_2 \sin \omega t$ and we assume $x_p = At \cos \omega t + Bt \sin \omega t$. Substituting into the DE we find $A = -F_0/2\omega$ and $B = 0$. Thus $x = c_1 \cos \omega t + c_2 \sin \omega t - (F_0/2\omega)t \cos \omega t$. From the initial conditions we obtain $c_1 = 0$ and $c_2 = F_0/2\omega^2$, so

$$x = (F_0/2\omega^2) \sin \omega t - (F_0/2\omega)t \cos \omega t.$$

34. (*Suggestion*) Note that in Problem 33, $\gamma = \omega$, whereas in this problem we will assume that $\gamma \neq \omega$.

36. We have $y_c = c_1 e^{-2x} + e^x(c_2 \cos \sqrt{3}\,x + c_3 \sin \sqrt{3}\,x)$ and we assume $y_p = Ax + B + Cxe^{-2x}$. Substituting into the DE we find $A = \frac{1}{4}$, $B = -\frac{5}{8}$, and $C = \frac{2}{3}$. Thus

$$y = c_1 e^{-2x} + e^x(c_2 \cos \sqrt{3}\,x + c_3 \sin \sqrt{3}\,x) + \frac{1}{4}x - \frac{5}{8} + \frac{2}{3}xe^{-2x}.$$

From the initial conditions we obtain $c_1 = -\frac{23}{12}$, $c_2 = -\frac{59}{24}$, and $c_3 = \frac{17}{72}\sqrt{3}$, so

$$y = -\frac{23}{12}e^{-2x} + e^x\left(-\frac{59}{24}\cos \sqrt{3}\,x + \frac{17}{72}\sqrt{3}\sin \sqrt{3}\,x\right) + \frac{1}{4}x - \frac{5}{8} + \frac{2}{3}xe^{-2x}.$$

39. The general solution of the DE $y'' + 3y = 6x$ is $y = c_1 \cos \sqrt{3}x + c_2 \sin \sqrt{3}x + 2x$. The condition $y(0) = 0$ implies $c_1 = 0$ and so $y = c_2 \sin \sqrt{3}x + 2x$. The condition $y(1) + y'(1) = 0$ implies $c_2 \sin \sqrt{3} + 2 + c_2\sqrt{3} \cos \sqrt{3} + 2 = 0$ so $c_2 = -4/(\sin \sqrt{3} + \sqrt{3} \cos \sqrt{3})$. The solution is

$$y = \frac{-4 \sin \sqrt{3}x}{\sin \sqrt{3} + \sqrt{3} \cos \sqrt{3}} + 2x.$$

42. We have $y_c = e^x(c_1 \cos 3x + c_2 \sin 3x)$ and we assume $y_p = A$ on $[0, \pi]$. Substituting into the DE we find $A = 2$. Thus, $y = e^x(c_1 \cos 3x + c_2 \sin 3x) + 2$ on $[0, \pi]$. On (π, ∞) we have $y = e^x(c_3 \cos 3x + c_4 \sin 3x)$. From $y(0) = 0$ and $y'(0) = 0$ we obtain

$$c_1 = -2, \qquad c_1 + 3c_2 = 0.$$

Solving this system, we find $c_1 = -2$ and $c_2 = \frac{2}{3}$. Thus $y = e^x(-2\cos 3x + \frac{2}{3}\sin 3x) + 2$ on $[0, \pi]$. Now, continuity of y at $x = \pi$ implies

$$e^\pi\left(-2\cos 3\pi + \frac{2}{3}\sin 3\pi\right) + 2 = e^\pi(c_3\cos 3\pi + c_4\sin 3\pi)$$

or $2 + 2e^\pi = -c_3 e^\pi$ or $c_3 = -2e^{-\pi}(1 + e^\pi)$. Continuity of y' at π implies

$$\frac{20}{3}e^\pi \sin 3\pi = e^\pi[(c_3 + 3c_4)\cos 3\pi + (-3c_3 + c_4)\sin 3\pi]$$

or $-c_3 e^\pi - 3c_4 e^\pi = 0$. Since $c_3 = -2e^{-\pi}(1 + e^\pi)$ we have $c_4 = \frac{2}{3}e^{-\pi}(1 + e^\pi)$. The solution of the IVP is

$$y(x) = \begin{cases} e^x(-2\cos 3x + \frac{2}{3}\sin 3x) + 2, & 0 \le x \le \pi \\ (1 + e^\pi)e^{x-\pi}(-2\cos 3x + \frac{2}{3}\sin 3x), & x > \pi. \end{cases}$$

Section 4.5 Undetermined Coefficients – Annihilator Approach

The terminology and concepts listed below provide an outline of the main ideas encountered in this section. These can be useful when preparing for a quiz or test.

Terminology and Concepts

- annihilator operator
- method of undetermined coefficients

The basic skills listed below summarize the more mechanical types of problems encountered in the exercise set for this section.

Basic Skills

- write a DE using differential operator notation
- find a differential operator that annihilates a given function
- find linearly independent functions that are annihilated by a given differential operator
- solve a nonhomogeneous linear DE with constant coefficients using annihilators and undetermined coefficients

Exercises 4.5   *Hints, Suggestions, Solutions, and Examples*

3. $(D^2 - 4D - 12)y = (D - 6)(D + 2)y = x - 6$

6. $(D^3 + 4D)y = D(D^2 + 4)y = e^x \cos 2x$

9. $(D^4 + 8D)y = D(D + 2)(D^2 - 2D + 4)y = 4$

12. $(2D - 1)y = (2D - 1)4e^{x/2} = 8De^{x/2} - 4e^{x/2} = 4e^{x/2} - 4e^{x/2} = 0$

15. D^4 because of x^3

18. $D^2(D - 6)^2$ because of x and xe^{6x}

21. $D^3(D^2 + 16)$ because of x^2 and $\sin 4x$

24. $D(D - 1)(D - 2)$ because of 1, e^x, and e^{2x}

27. 1, x, x^2, x^3, x^4

30. $D^2 - 9D - 36 = (D - 12)(D + 3)$; e^{12x}, e^{-3x}

33. $D^3 - 10D^2 + 25D = D(D - 5)^2$; 1, e^{5x}, xe^{5x}

36. Applying D to the DE we obtain

$$D(2D^2 - 7D + 5)y = 0.$$

Then

$$y = \underbrace{c_1 e^{5x/2} + c_2 e^x}_{y_c} + c_3$$

and $y_p = A$. Substituting y_p into the DE yields $5A = -29$ or $A = -29/5$. The general solution is

$$y = c_1 e^{5x/2} + c_2 e^x - \frac{29}{5}.$$

39. Applying D^2 to the DE we obtain

$$D^2(D^2 + 4D + 4)y = D^2(D + 2)^2 y = 0.$$

Then

$$y = \underbrace{c_1 e^{-2x} + c_2 x e^{-2x}}_{y_c} + c_3 + c_4 x$$

and $y_p = Ax + B$. Substituting y_p into the DE yields $4Ax + (4A + 4B) = 2x + 6$. Equating coefficients gives

$$4A = 2$$

$$4A + 4B = 6.$$

Then $A = 1/2$, $B = 1$, and the general solution is

$$y = c_1 e^{-2x} + c_2 x e^{-2x} + \frac{1}{2}x + 1.$$

42. Applying D^4 to the DE we obtain

$$D^4(D^2 - 2D + 1)y = D^4(D-1)^2 y = 0.$$

Then

$$y = \underbrace{c_1 e^x + c_2 x e^x}_{y_c} + c_3 x^3 + c_4 x^2 + c_5 x + c_6$$

and $y_p = Ax^3 + Bx^2 + Cx + E$. Substituting y_p into the DE yields

$$Ax^3 + (B - 6A)x^2 + (6A - 4B + C)x + (2B - 2C + E) = x^3 + 4x.$$

Equating coefficients gives

$$A = 1$$

$$B - 6A = 0$$

$$6A - 4B + C = 4$$

$$2B - 2C + E = 0.$$

Then $A = 1$, $B = 6$, $C = 22$, $E = 32$, and the general solution is

$$y = c_1 e^x + c_2 x e^x + x^3 + 6x^2 + 22x + 32.$$

45. Applying $D(D-1)$ to the DE we obtain

$$D(D-1)(D^2 - 2D - 3)y = D(D-1)(D+1)(D-3)y = 0.$$

Then

$$y = \underbrace{c_1 e^{3x} + c_2 e^{-x}}_{y_c} + c_3 e^x + c_4$$

and $y_p = Ae^x + B$. Substituting y_p into the DE yields $-4Ae^x - 3B = 4e^x - 9$. Equating coefficients gives $A = -1$ and $B = 3$. The general solution is

$$y = c_1 e^{3x} + c_2 e^{-x} - e^x + 3.$$

48. Applying $D(D^2 + 1)$ to the DE we obtain

$$D(D^2 + 1)(D^2 + 4)y = 0.$$

Then

$$y = \underbrace{c_1 \cos 2x + c_2 \sin 2x}_{y_c} + c_3 \cos x + c_4 \sin x + c_5$$

and $y_p = A \cos x + B \sin x + C$. Substituting y_p into the DE yields

$$3A \cos x + 3B \sin x + 4C = 4 \cos x + 3 \sin x - 8.$$

Equating coefficients gives $A = 4/3$, $B = 1$, and $C = -2$. The general solution is

$$y = c_1 \cos 2x + c_2 \sin 2x + \frac{4}{3} \cos x + \sin x - 2.$$

51. Applying $D(D-1)^3$ to the DE we obtain

$$D(D-1)^3(D^2-1)y = D(D-1)^4(D+1)y = 0.$$

Then

$$y = \underbrace{c_1 e^x + c_2 e^{-x}}_{y_c} + c_3 x^3 e^x + c_4 x^2 e^x + c_5 x e^x + c_6$$

and $y_p = Ax^3 e^x + Bx^2 e^x + Cx e^x + E$. Substituting y_p into the DE yields

$$6Ax^2 e^x + (6A + 4B)x e^x + (2B + 2C)e^x - E = x^2 e^x + 5.$$

Equating coefficients gives

$$6A = 1$$

$$6A + 4B = 0$$

$$2B + 2C = 0$$

$$-E = 5.$$

Then $A = 1/6$, $B = -1/4$, $C = 1/4$, $E = -5$, and the general solution is

$$y = c_1 e^x + c_2 e^{-x} + \frac{1}{6}x^3 e^x - \frac{1}{4}x^2 e^x + \frac{1}{4}x e^x - 5.$$

54. Applying $D^2 - 2D + 10$ to the DE we obtain

$$(D^2 - 2D + 10)\left(D^2 + D + \frac{1}{4}\right)y = (D^2 - 2D + 10)\left(D + \frac{1}{2}\right)^2 y = 0.$$

Then

$$y = \underbrace{c_1 e^{-x/2} + c_2 x e^{-x/2}}_{y_c} + c_3 e^x \cos 3x + c_4 e^x \sin 3x$$

and $y_p = Ae^x \cos 3x + Be^x \sin 3x$. Substituting y_p into the DE yields

$$(9B - 27A/4)e^x \cos 3x - (9A + 27B/4)e^x \sin 3x = -e^x \cos 3x + e^x \sin 3x.$$

Equating coefficients gives

$$-\frac{27}{4}A + 9B = -1$$

$$-9A - \frac{27}{4}B = 1.$$

Then $A = -4/225$, $B = -28/225$, and the general solution is

$$y = c_1 e^{-x/2} + c_2 x e^{-x/2} - \frac{4}{225}e^x \cos 3x - \frac{28}{225}e^x \sin 3x.$$

57. Applying $(D^2 + 1)^2$ to the DE we obtain

$$(D^2 + 1)^2(D^2 + D + 1) = 0.$$

Then

$$y = e^{-x/2}\underbrace{\left[c_1 \cos \frac{\sqrt{3}}{2} x + c_2 \sin \frac{\sqrt{3}}{2} x\right] + c_3 \cos x + c_4 \sin x + c_5 x \cos x + c_6 x \sin x}_{y_c}$$

and $y_p = A \cos x + B \sin x + C x \cos x + E x \sin x$. Substituting y_p into the DE yields

$$(B + C + 2E) \cos x + E x \cos x + (-A - 2C + E) \sin x - C x \sin x = x \sin x.$$

Equating coefficients gives

$$B + C + 2E = 0$$

$$E = 0$$

$$-A - 2C + E = 0$$

$$-C = 1.$$

Then $A = 2$, $B = 1$, $C = -1$, and $E = 0$, and the general solution is

$$y = e^{-x/2}\left[c_1 \cos \frac{\sqrt{3}}{2} x + c_2 \sin \frac{\sqrt{3}}{2} x\right] + 2 \cos x + \sin x - x \cos x.$$

60. Applying $D(D - 1)^2(D + 1)$ to the DE we obtain

$$D(D - 1)^2(D + 1)(D^3 - D^2 + D - 1) = D(D - 1)^3(D + 1)(D^2 + 1) = 0.$$

Then

$$y = \underbrace{c_1 e^x + c_2 \cos x + c_3 \sin x}_{y_c} + c_4 + c_5 e^{-x} + c_6 x e^x + c_7 x^2 e^x$$

and $y_p = A + B e^{-x} + C x e^x + E x^2 e^x$. Substituting y_p into the DE yields

$$4 E x e^x + (2C + 4E) e^x - 4 B e^{-x} - A = x e^x - e^{-x} + 7.$$

Equating coefficients gives

$$4E = 1$$

$$2C + 4E = 0$$

$$-4B = -1$$

$$-A = 7.$$

Then $A = -7$, $B = 1/4$, $C = -1/2$, and $E = 1/4$, and the general solution is

$$y = c_1 e^x + c_2 \cos x + c_3 \sin x - 7 + \frac{1}{4} e^{-x} - \frac{1}{2} x e^x + \frac{1}{4} x^2 e^x.$$

63. Applying $D(D-1)$ to the DE we obtain

$$D(D-1)(D^4 - 2D^3 + D^2) = D^3(D-1)^3 = 0.$$

Then

$$y = \underbrace{c_1 + c_2 x + c_3 e^x + c_4 x e^x}_{y_c} + c_5 x^2 + c_6 x^2 e^x$$

and $y_p = Ax^2 + Bx^2 e^x$. Substituting y_p into the DE yields $2A + 2Be^x = 1 + e^x$. Equating coefficients gives $A = 1/2$ and $B = 1/2$. The general solution is

$$y = c_1 + c_2 x + c_3 e^x + c_4 x e^x + \frac{1}{2}x^2 + \frac{1}{2}x^2 e^x.$$

66. The complementary function is $y_c = c_1 + c_2 e^{-x}$. Using D^2 to annihilate x we find $y_p = Ax + Bx^2$. Substituting y_p into the DE we obtain $(A + 2B) + 2Bx = x$. Thus $A = -1$ and $B = 1/2$, and

$$y = c_1 + c_2 e^{-x} - x + \frac{1}{2}x^2$$

$$y' = -c_2 e^{-x} - 1 + x.$$

The initial conditions imply

$$c_1 + c_2 = 1$$

$$-c_2 = 1.$$

Thus $c_1 = 2$ and $c_2 = -1$, and

$$y = 2 - e^{-x} - x + \frac{1}{2}x^2.$$

69. The complementary function is $y_c = c_1 \cos x + c_2 \sin x$. Using $(D^2 + 1)(D^2 + 4)$ to annihilate $8 \cos 2x - 4 \sin x$ we find $y_p = Ax \cos x + Bx \sin x + C \cos 2x + E \sin 2x$. Substituting y_p into the DE we obtain $2B \cos x - 3C \cos 2x - 2A \sin x - 3E \sin 2x = 8 \cos 2x - 4 \sin x$. Thus $A = 2$, $B = 0$, $C = -8/3$, and $E = 0$, and

$$y = c_1 \cos x + c_2 \sin x + 2x \cos x - \frac{8}{3}\cos 2x$$

$$y' = -c_1 \sin x + c_2 \cos x + 2 \cos x - 2x \sin x + \frac{16}{3}\sin 2x.$$

The initial conditions imply

$$c_2 + \frac{8}{3} = -1$$

$$-c_1 - \pi = 0.$$

Thus $c_1 = -\pi$ and $c_2 = -11/3$, and

$$y = -\pi \cos x - \frac{11}{3}\sin x + 2x \cos x - \frac{8}{3}\cos 2x.$$

72. The complementary function is $y_c = c_1 + c_2 x + c_3 x^2 + c_4 e^x$. Using $D^2(D-1)$ to annihilate $x + e^x$ we find $y_p = Ax^3 + Bx^4 + Cxe^x$. Substituting y_p into the DE we obtain $(-6A + 24B) - 24Bx + Ce^x = x + e^x$. Thus $A = -1/6$, $B = -1/24$, and $C = 1$, and

$$y = c_1 + c_2 x + c_3 x^2 + c_4 e^x - \frac{1}{6}x^3 - \frac{1}{24}x^4 + xe^x$$

$$y' = c_2 + 2c_3 x + c_4 e^x - \frac{1}{2}x^2 - \frac{1}{6}x^3 + e^x + xe^x$$

$$y'' = 2c_3 + c_4 e^x - x - \frac{1}{2}x^2 + 2e^x + xe^x$$

$$y''' = c_4 e^x - 1 - x + 3e^x + xe^x.$$

The initial conditions imply

$$c_1 + c_4 = 0$$

$$c_2 + c_4 + 1 = 0$$

$$2c_3 + c_4 + 2 = 0$$

$$2 + c_4 = 0.$$

Thus $c_1 = 2$, $c_2 = 1$, $c_3 = 0$, and $c_4 = -2$, and

$$y = 2 + x - 2e^x - \frac{1}{6}x^3 - \frac{1}{24}x^4 + xe^x.$$

Section 4.6 Variation of Parameters

The terminology and concepts listed below provide an outline of the main ideas encountered in this section. These can be useful when preparing for a quiz or test.

Terminology and Concepts

- variation of parameters

The basic skills listed below summarize the more mechanical types of problems encountered in the exercise set for this section.

Basic Skills

- solve a nonhomogeneous linear second-order DE with constant coefficients using variation of parameters

Integrals of Trig Functions In Example 2 in the text we had to integrate

$$\int \cot 3x \, dx = \int \frac{\cos 3x}{\sin 3x} \, dx.$$

If we identify $u = \sin 3x$, then $du = 3\cos 3x \, dx$. The integral is recognized as one of the form $\int du/u = \ln |u| + c$:

$$\frac{1}{3} \int \frac{\overbrace{3\cos 3x \, dx}^{du}}{\underbrace{\sin 3x}_{u}} = \frac{1}{3} \ln |\sin 3x| + c.$$

In general, we can integrate $\tan u$ and $\cot u$ in terms of logarithms:

$$\int \tan u \, du = -\ln |\cos u| + c \tag{1}$$

$$\int \cot u \, du = \ln |\sin u| + c. \tag{2}$$

The integrals in (1) and (2), along with

$$\int \sec u \, du = \ln |\sec u + \tan u| + c \tag{3}$$

$$\int \csc u \, du = \ln |\csc u - \cot u| + c \tag{4}$$

occur frequently in the solution of differential equations by variation of parameters. Watch for them in Exercises 4.6.

Exercises 4.6 *Hints, Suggestions, Solutions, and Examples*

The particular solution, $y_p = u_1 y_1 + u_2 y_2$, in the following problems can take on a variety of forms, especially where trigonometric functions are involved. The validity of a particular form can best be checked by substituting it back into the differential equation.

3. The auxiliary equation is $m^2 + 1 = 0$, so $y_c = c_1 \cos x + c_2 \sin x$ and

$$W = \begin{vmatrix} \cos x & \sin x \\ -\sin x & \cos x \end{vmatrix} = 1.$$

Identifying $f(x) = \sin x$ we obtain

$$u_1' = -\sin^2 x$$

$$u_2' = \cos x \sin x.$$

Then

$$u_1 = \frac{1}{4}\sin 2x - \frac{1}{2}x = \frac{1}{2}\sin x \cos x - \frac{1}{2}x$$

$$u_2 = -\frac{1}{2}\cos^2 x,$$

and

$$y = c_1 \cos x + c_2 \sin x + \frac{1}{2} \sin x \cos^2 x - \frac{1}{2} x \cos x - \frac{1}{2} \cos^2 x \sin x$$

$$= c_1 \cos x + c_2 \sin x - \frac{1}{2} x \cos x.$$

5. (*Hint*) To integrate $\sin x \cos^2 x$ think of $\int u^2 \, du$ with $u = \cos x$ and $du = -\sin x \, dx$.

6. The auxiliary equation is $m^2 + 1 = 0$, so $y_c = c_1 \cos x + c_2 \sin x$ and

$$W = \begin{vmatrix} \cos x & \sin x \\ -\sin x & \cos x \end{vmatrix} = 1.$$

Identifying $f(x) = \sec^2 x$ we obtain

$$u_1' = -\frac{\sin x}{\cos^2 x}$$

$$u_2' = \sec x.$$

Then

$$u_1 = -\frac{1}{\cos x} = -\sec x$$

$$u_2 = \ln|\sec x + \tan x|$$

and

$$y = c_1 \cos x + c_2 \sin x - \cos x \sec x + \sin x \ln|\sec x + \tan x|$$

$$= c_1 \cos x + c_2 \sin x - 1 + \sin x \ln|\sec x + \tan x|.$$

9. The auxiliary equation is $m^2 - 4 = 0$, so $y_c = c_1 e^{2x} + c_2 e^{-2x}$ and

$$W = \begin{vmatrix} e^{2x} & e^{-2x} \\ 2e^{2x} & -2e^{-2x} \end{vmatrix} = -4.$$

Identifying $f(x) = e^{2x}/x$ we obtain $u_1' = 1/4x$ and $u_2' = -e^{4x}/4x$. Then

$$u_1 = \frac{1}{4} \ln|x|,$$

$$u_2 = -\frac{1}{4} \int_{x_0}^x \frac{e^{4t}}{t} \, dt$$

and

$$y = c_1 e^{2x} + c_2 e^{-2x} + \frac{1}{4}\left(e^{2x} \ln|x| - e^{-2x} \int_{x_0}^x \frac{e^{4t}}{t} \, dt \right), \qquad x_0 > 0.$$

12. The auxiliary equation is $m^2 - 2m + 1 = (m-1)^2 = 0$, so $y_c = c_1 e^x + c_2 x e^x$ and

$$W = \begin{vmatrix} e^x & x e^x \\ e^x & x e^x + e^x \end{vmatrix} = e^{2x}.$$

Identifying $f(x) = e^x / \left(1 + x^2\right)$ we obtain

$$u_1' = -\frac{xe^x e^x}{e^{2x}\left(1 + x^2\right)} = -\frac{x}{1 + x^2}$$

$$u_2' = \frac{e^x e^x}{e^{2x}\left(1 + x^2\right)} = \frac{1}{1 + x^2}.$$

Then $u_1 = -\frac{1}{2}\ln\left(1 + x^2\right)$, $u_2 = \tan^{-1} x$, and

$$y = c_1 e^x + c_2 x e^x - \frac{1}{2}e^x \ln\left(1 + x^2\right) + xe^x \tan^{-1} x.$$

13. (*Hint*) To integrate $e^{2x} \sin e^x$ think of the form $\int u \sin u \, du$, with $u = e^x$, $du = e^x \, dx$, and then use integration by parts.

15. The auxiliary equation is $m^2 + 2m + 1 = (m+1)^2 = 0$, so $y_c = c_1 e^{-t} + c_2 t e^{-t}$ and

$$W = \begin{vmatrix} e^{-t} & te^{-t} \\ -e^{-t} & -te^{-t} + e^{-t} \end{vmatrix} = e^{-2t}.$$

Identifying $f(t) = e^{-t} \ln t$ we obtain

$$u_1' = -\frac{te^{-t}e^{-t}\ln t}{e^{-2t}} = -t \ln t$$

$$u_2' = \frac{e^{-t}e^{-t}\ln t}{e^{-2t}} = \ln t.$$

Then

$$u_1 = -\frac{1}{2}t^2 \ln t + \frac{1}{4}t^2$$

$$u_2 = t \ln t - t$$

and

$$y = c_1 e^{-t} + c_2 t e^{-t} - \frac{1}{2}t^2 e^{-t} \ln t + \frac{1}{4}t^2 e^{-t} + t^2 e^{-t} \ln t - t^2 e^{-t}$$

$$= c_1 e^{-t} + c_2 t e^{-t} + \frac{1}{2}t^2 e^{-t} \ln t - \frac{3}{4}t^2 e^{-t}.$$

18. The auxiliary equation is $4m^2 - 4m + 1 = (2m-1)^2 = 0$, so $y_c = c_1 e^{x/2} + c_2 x e^{x/2}$ and

$$W = \begin{vmatrix} e^{x/2} & xe^{x/2} \\ \frac{1}{2}e^{x/2} & \frac{1}{2}xe^{x/2} + e^{x/2} \end{vmatrix} = e^x.$$

Identifying $f(x) = \frac{1}{4}e^{x/2}\sqrt{1 - x^2}$ we obtain

$$u_1' = -\frac{xe^{x/2}e^{x/2}\sqrt{1 - x^2}}{4e^x} = -\frac{1}{4}x\sqrt{1 - x^2}$$

$$u_2' = \frac{e^{x/2}e^{x/2}\sqrt{1 - x^2}}{4e^x} = \frac{1}{4}\sqrt{1 - x^2}.$$

To find u_1 and u_2 we use the substitution $v = 1 - x^2$ and the trig substitution $x = \sin\theta$, respectively:

$$u_1 = \frac{1}{12}\left(1 - x^2\right)^{3/2}$$

$$u_2 = \frac{x}{8}\sqrt{1 - x^2} + \frac{1}{8}\sin^{-1}x.$$

Thus

$$y = c_1 e^{x/2} + c_2 x e^{x/2} + \frac{1}{12}e^{x/2}\left(1 - x^2\right)^{3/2} + \frac{1}{8}x^2 e^{x/2}\sqrt{1 - x^2} + \frac{1}{8}x e^{x/2}\sin^{-1}x.$$

21. The auxiliary equation is $m^2 + 2m - 8 = (m - 2)(m + 4) = 0$, so $y_c = c_1 e^{2x} + c_2 e^{-4x}$ and

$$W = \begin{vmatrix} e^{2x} & e^{-4x} \\ 2e^{2x} & -4e^{-4x} \end{vmatrix} = -6e^{-2x}.$$

Identifying $f(x) = 2e^{-2x} - e^{-x}$ we obtain

$$u_1' = \frac{1}{3}e^{-4x} - \frac{1}{6}e^{-3x}$$

$$u_2' = \frac{1}{6}e^{3x} - \frac{1}{3}e^{2x}.$$

Then

$$u_1 = -\frac{1}{12}e^{-4x} + \frac{1}{18}e^{-3x}$$

$$u_2 = \frac{1}{18}e^{3x} - \frac{1}{6}e^{2x}.$$

Thus

$$y = c_1 e^{2x} + c_2 e^{-4x} - \frac{1}{12}e^{-2x} + \frac{1}{18}e^{-x} + \frac{1}{18}e^{-x} - \frac{1}{6}e^{-2x}$$

$$= c_1 e^{2x} + c_2 e^{-4x} - \frac{1}{4}e^{-2x} + \frac{1}{9}e^{-x}$$

and

$$y' = 2c_1 e^{2x} - 4c_2 e^{-4x} + \frac{1}{2}e^{-2x} - \frac{1}{9}e^{-x}.$$

The initial conditions imply

$$c_1 + \ c_2 - \frac{5}{36} = 1$$

$$2c_1 - 4c_2 + \frac{7}{18} = 0.$$

Thus $c_1 = 25/36$ and $c_2 = 4/9$, and

$$y = \frac{25}{36}e^{2x} + \frac{4}{9}e^{-4x} - \frac{1}{4}e^{-2x} + \frac{1}{9}e^{-x}.$$

24. Write the equation in the form

$$y'' + \frac{1}{x}y' + \frac{1}{x^2}y = \frac{\sec(\ln x)}{x^2}$$

and identify $f(x) = \sec(\ln x)/x^2$. From $y_1 = \cos(\ln x)$ and $y_2 = \sin(\ln x)$ we compute

$$W = \begin{vmatrix} \cos(\ln x) & \sin(\ln x) \\ -\dfrac{\sin(\ln x)}{x} & \dfrac{\cos(\ln x)}{x} \end{vmatrix} = \dfrac{1}{x}.$$

Now

$$u_1' = -\frac{\tan(\ln x)}{x} \quad \text{so} \quad u_1 = \ln|\cos(\ln x)|,$$

and

$$u_2' = \frac{1}{x} \quad \text{so} \quad u_2 = \ln x.$$

Thus, a particular solution is

$$y_p = \cos(\ln x)\ln|\cos(\ln x)| + (\ln x)\sin(\ln x),$$

and the general solution is

$$y = c_1\cos(\ln x) + c_2\sin(\ln x) + \cos(\ln x)\ln|\cos(\ln x)| + (\ln x)\sin(\ln x).$$

25-26. (*Suggestion*) A rule for the evaluation of determinants that can be very useful here is that a determinant can be expanded by any row or any column. Thus, if a determinant has a row (or column) that contains many zeros, it is easiest to expand the determinant by that row (or column).

Section 4.7   Cauchy-Euler Equation

The terminology and concepts listed below provide an outline of the main ideas encountered in this section. These can be useful when preparing for a quiz or test.

Terminology and Concepts

- Cauchy-Euler equation
- auxiliary equation
- variation of parameters

The basic skills listed below summarize the more mechanical types of problems encountered in the exercise set for this section.

Basic Skills

- find the general solution of a homogeneous Cauchy-Euler equation

- use variation of parameters to find a particular solution of a nonhomogeneous Cauchy-Euler equation

- use the substitution $x = e^t$ to transform a Cauchy-Euler equation into a linear DE with constant coefficients

Exercises 4.7   *Hints, Suggestions, Solutions, and Examples*

3. The auxiliary equation is $m^2 = 0$ so that $y = c_1 + c_2 \ln x$.

6. The auxiliary equation is $m^2 + 4m + 3 = (m+1)(m+3) = 0$ so that $y = c_1 x^{-1} + c_2 x^{-3}$.

9. The auxiliary equation is $25m^2 + 1 = 0$ so that $y = c_1 \cos\left(\frac{1}{5}\ln x\right) + c_2 \sin\left(\frac{1}{5}\ln x\right)$.

12. The auxiliary equation is $m^2 + 7m + 6 = (m+1)(m+6) = 0$ so that $y = c_1 x^{-1} + c_2 x^{-6}$.

15. Assuming that $y = x^m$ and substituting into the DE we obtain

$$m(m-1)(m-2) - 6 = m^3 - 3m^2 + 2m - 6 = (m-3)(m^2+2) = 0.$$

Thus

$$y = c_1 x^3 + c_2 \cos\left(\sqrt{2}\ln x\right) + c_3 \sin\left(\sqrt{2}\ln x\right).$$

18. Assuming that $y = x^m$ and substituting into the DE we obtain

$$m(m-1)(m-2)(m-3) + 6m(m-1)(m-2) + 9m(m-1) + 3m + 1 = m^4 + 2m^2 + 1 = (m^2+1)^2 = 0.$$

Thus

$$y = c_1 \cos(\ln x) + c_2 \sin(\ln x) + c_3(\ln x)\cos(\ln x) + c_4(\ln x)\sin(\ln x).$$

21. The auxiliary equation is $m^2 - 2m + 1 = (m-1)^2 = 0$ so that $y_c = c_1 x + c_2 x \ln x$ and

$$W(x, x\ln x) = \begin{vmatrix} x & x\ln x \\ 1 & 1 + \ln x \end{vmatrix} = x.$$

Identifying $f(x) = 2/x$ we obtain $u_1' = -2\ln x/x$ and $u_2' = 2/x$. Then $u_1 = -(\ln x)^2$, $u_2 = 2\ln x$, and

$$y = c_1 x + c_2 x \ln x - x(\ln x)^2 + 2x(\ln x)^2$$

$$= c_1 x + c_2 x \ln x + x(\ln x)^2, \qquad x > 0.$$

24. The auxiliary equation $m(m-1) + m - 1 = m^2 - 1 = 0$ has roots $m_1 = -1$, $m_2 = 1$, so $y_c = c_1 x^{-1} + c_2 x$. With $y_1 = x^{-1}$, $y_2 = x$, and the identification $f(x) = 1/x^2(x+1)$, we get

$$W = 2x^{-1}, \qquad W_1 = -1/x(x+1), \qquad \text{and} \qquad W_2 = 1/x^3(x+1).$$

Then $u_1' = W_1/W = -1/2(x+1),$ $u_2' = W_2/W = 1/2x^2(x+1),$ and integration (by partial fractions for u_2') gives

$$u_1 = -\frac{1}{2}\ln(x+1)$$

$$u_2 = -\frac{1}{2}x^{-1} - \frac{1}{2}\ln x + \frac{1}{2}\ln(x+1),$$

so

$$y_p = u_1 y_1 + u_2 y_2 = \left[-\frac{1}{2}\ln(x+1)\right]x^{-1} + \left[-\frac{1}{2}x^{-1} - \frac{1}{2}\ln x + \frac{1}{2}\ln(x+1)\right]x$$

$$= -\frac{1}{2} - \frac{1}{2}x\ln x + \frac{1}{2}x\ln(x+1) - \frac{\ln(x+1)}{2x} = -\frac{1}{2} + \frac{1}{2}x\ln\left(1+\frac{1}{x}\right) - \frac{\ln(x+1)}{2x}$$

and

$$y = y_c + y_p = c_1 x^{-1} + c_2 x - \frac{1}{2} + \frac{1}{2}x\ln\left(1+\frac{1}{x}\right) - \frac{\ln(x+1)}{2x}, \qquad x > 0.$$

27. The auxiliary equation is $m^2 + 1 = 0$, so that

$$y = c_1\cos(\ln x) + c_2\sin(\ln x)$$

and

$$y' = -c_1\frac{1}{x}\sin(\ln x) + c_2\frac{1}{x}\cos(\ln x).$$

The initial conditions imply $c_1 = 1$ and $c_2 = 2$. Thus
$y = \cos(\ln x) + 2\sin(\ln x)$. The graph is given to the right.

30. The auxiliary equation is $m^2 - 6m + 8 = (m-2)(m-4) = 0$, so
that $y_c = c_1 x^2 + c_2 x^4$ and

$$W = \begin{vmatrix} x^2 & x^4 \\ 2x & 4x^3 \end{vmatrix} = 2x^5.$$

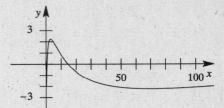

Identifying $f(x) = 8x^4$ we obtain $u_1' = -4x^3$ and $u_2' = 4x$. Then
$u_1 = -x^4$, $u_2 = 2x^2$, and $y = c_1 x^2 + c_2 x^4 + x^6$. The initial conditions imply

$$\frac{1}{4}c_1 + \frac{1}{16}c_2 = -\frac{1}{64}$$

$$c_1 + \frac{1}{2}c_2 = -\frac{3}{16}.$$

Thus $c_1 = \frac{1}{16}$, $c_2 = -\frac{1}{2}$, and $y = \frac{1}{16}x^2 - \frac{1}{2}x^4 + x^6$. The graph is given above.

33. Substituting $x = e^t$ into the DE we obtain

$$\frac{d^2y}{dt^2} + 9\frac{dy}{dt} + 8y = e^{2t}.$$

The auxiliary equation is $m^2 + 9m + 8 = (m+1)(m+8) = 0$ so that $y_c = c_1 e^{-t} + c_2 e^{-8t}$. Using undetermined coefficients we try $y_p = Ae^{2t}$. This leads to $30Ae^{2t} = e^{2t}$, so that $A = 1/30$ and

$$y = c_1 e^{-t} + c_2 e^{-8t} + \frac{1}{30} e^{2t} = c_1 x^{-1} + c_2 x^{-8} + \frac{1}{30} x^2.$$

36. From

$$\frac{d^2 y}{dx^2} = \frac{1}{x^2} \left(\frac{d^2 y}{dt^2} - \frac{dy}{dt} \right)$$

it follows that

$$\frac{d^3 y}{dx^3} = \frac{1}{x^2} \frac{d}{dx} \left(\frac{d^2 y}{dt^2} - \frac{dy}{dt} \right) - \frac{2}{x^3} \left(\frac{d^2 y}{dt^2} - \frac{dy}{dt} \right)$$

$$= \frac{1}{x^2} \frac{d}{dx} \left(\frac{d^2 y}{dt^2} \right) - \frac{1}{x^2} \frac{d}{dx} \left(\frac{dy}{dt} \right) - \frac{2}{x^3} \frac{d^2 y}{dt^2} + \frac{2}{x^3} \frac{dy}{dt}$$

$$= \frac{1}{x^2} \frac{d^3 y}{dt^3} \left(\frac{1}{x} \right) - \frac{1}{x^2} \frac{d^2 y}{dt^2} \left(\frac{1}{x} \right) - \frac{2}{x^3} \frac{d^2 y}{dt^2} + \frac{2}{x^3} \frac{dy}{dt}$$

$$= \frac{1}{x^3} \left(\frac{d^3 y}{dt^3} - 3 \frac{d^2 y}{dt^2} + 2 \frac{dy}{dt} \right).$$

Substituting into the DE we obtain

$$\frac{d^3 y}{dt^3} - 3 \frac{d^2 y}{dt^2} + 2 \frac{dy}{dt} - 3 \left(\frac{d^2 y}{dt^2} - \frac{dy}{dt} \right) + 6 \frac{dy}{dt} - 6y = 3 + 3t$$

or

$$\frac{d^3 y}{dt^3} - 6 \frac{d^2 y}{dt^2} + 11 \frac{dy}{dt} - 6y = 3 + 3t.$$

The auxiliary equation is $m^3 - 6m^2 + 11m - 6 = (m-1)(m-2)(m-3) = 0$ so that $y_c = c_1 e^t + c_2 e^{2t} + c_3 e^{3t}$. Using undetermined coefficients we try $y_p = A + Bt$. This leads to $(11B - 6A) - 6Bt = 3 + 3t$, so that $A = -17/12$, $B = -1/2$, and

$$y = c_1 e^t + c_2 e^{2t} + c_3 e^{3t} - \frac{17}{12} - \frac{1}{2} t = c_1 x + c_2 x^2 + c_3 x^3 - \frac{17}{12} - \frac{1}{2} \ln x.$$

37-38. Use the substitution $t = -x$ since the initial conditions are on the interval $(-\infty, 0)$. See Example 6 in the text for an example of how to find dy/dt and $d^2 y/dt^2$.

39. (*Hint*) Think substitution.

Section 4.8

Solving Systems of Linear DEs by Elimination

The terminology and concepts listed below provide an outline of the main ideas encountered in this section. These can be useful when preparing for a quiz or test.

Terminology and Concepts

- system of linear DEs

- systematic elimination to find the solution of a system of linear DEs

The basic skills listed below summarize the more mechanical types of problems encountered in the exercise set for this section.

Basic Skills

- use systematic elimnation to solve a system of linear DEs

Exercises 4.8   *Hints, Suggestions, Solutions, and Examples*

3. From $Dx = -y + t$ and $Dy = x - t$ we obtain $y = t - Dx$, $Dy = 1 - D^2x$, and $(D^2 + 1)x = 1 + t$. The solution is

$$x = c_1 \cos t + c_2 \sin t + 1 + t$$

$$y = c_1 \sin t - c_2 \cos t + t - 1.$$

6. From $(D + 1)x + (D - 1)y = 2$ and $3x + (D + 2)y = -1$ we obtain $x = -\frac{1}{3} - \frac{1}{3}(D + 2)y$, $Dx = -\frac{1}{3}(D^2 + 2D)y$, and $(D^2 + 5)y = -7$. The solution is

$$y = c_1 \cos \sqrt{5}\, t + c_2 \sin \sqrt{5}\, t - \frac{7}{5}$$

$$x = \left(-\frac{2}{3}c_1 - \frac{\sqrt{5}}{3}c_2\right) \cos \sqrt{5}\, t + \left(\frac{\sqrt{5}}{3}c_1 - \frac{2}{3}c_2\right) \sin \sqrt{5}\, t + \frac{3}{5}.$$

9. From $Dx + D^2y = e^{3t}$ and $(D + 1)x + (D - 1)y = 4e^{3t}$ we obtain $D(D^2 + 1)x = 34e^{3t}$ and $D(D^2 + 1)y = -8e^{3t}$. The solution is

$$y = c_1 + c_2 \sin t + c_3 \cos t - \frac{4}{15}e^{3t}$$

$$x = c_4 + c_5 \sin t + c_6 \cos t + \frac{17}{15}e^{3t}.$$

98

Substituting into $(D+1)x + (D-1)y = 4e^{3t}$ gives

$$(c_4 - c_1) + (c_5 - c_6 - c_3 - c_2)\sin t + (c_6 + c_5 + c_2 - c_3)\cos t = 0$$

so that $c_4 = c_1$, $c_5 = c_3$, $c_6 = -c_2$, and

$$x = c_1 - c_2\cos t + c_3\sin t + \frac{17}{15}e^{3t}.$$

12. From $(2D^2 - D - 1)x - (2D+1)y = 1$ and $(D-1)x + Dy = -1$ we obtain $(2D+1)(D-1)(D+1)x = -1$ and $(2D+1)(D+1)y = -2$. The solution is

$$x = c_1 e^{-t/2} + c_2 e^{-t} + c_3 e^t + 1$$

$$y = c_4 e^{-t/2} + c_5 e^{-t} - 2.$$

Substituting into $(D-1)x + Dy = -1$ gives

$$\left(-\frac{3}{2}c_1 - \frac{1}{2}c_4\right)e^{-t/2} + (-2c_2 - c_5)e^{-t} = 0$$

so that $c_4 = -3c_1$, $c_5 = -2c_2$, and

$$y = -3c_1 e^{-t/2} - 2c_2 e^{-t} - 2.$$

15. Multiplying the first equation by $D+1$ and the second equation by $D^2 + 1$ and subtracting we obtain $(D^4 - D^2)x = 1$. Then

$$x = c_1 + c_2 t + c_3 e^t + c_4 e^{-t} - \frac{1}{2}t^2.$$

Multiplying the first equation by $D+1$ and subtracting we obtain $D^2(D+1)y = 1$. Then

$$y = c_5 + c_6 t + c_7 e^{-t} - \frac{1}{2}t^2.$$

Substituting into $(D-1)x + (D^2+1)y = 1$ gives

$$(-c_1 + c_2 + c_5 - 1) + (-2c_4 + 2c_7)e^{-t} + (-1 - c_2 + c_6)t = 1$$

so that $c_5 = c_1 - c_2 + 2$, $c_6 = c_2 + 1$, and $c_7 = c_4$. The solution of the system is

$$x = c_1 + c_2 t + c_3 e^t + c_4 e^{-t} - \frac{1}{2}t^2$$

$$y = (c_1 - c_2 + 2) + (c_2 + 1)t + c_4 e^{-t} - \frac{1}{2}t^2.$$

18. From $Dx + z = e^t$, $(D-1)x + Dy + Dz = 0$, and $x + 2y + Dz = e^t$ we obtain $z = -Dx + e^t$, $Dz = -D^2 x + e^t$, and the system $(-D^2 + D - 1)x + Dy = -e^t$ and $(-D^2 + 1)x + 2y = 0$. Then $y = \frac{1}{2}(D^2 - 1)x$, $Dy = \frac{1}{2}D(D^2 - 1)x$, and $(D-2)(D^2 + 1)x = -2e^t$ so that the solution is

$$x = c_1 e^{2t} + c_2 \cos t + c_3 \sin t + e^t$$

$$y = \frac{3}{2}c_1 e^{2t} - c_2 \cos t - c_3 \sin t$$

$$z = -2c_1 e^{2t} - c_3 \cos t + c_2 \sin t.$$

21. From $(D+5)x+y = 0$ and $4x - (D+1)y = 0$ we obtain $y = -(D+5)x$ so that $Dy = -(D^2+5D)x$. Then $4x + (D^2 + 5D)x + (D+5)x = 0$ and $(D+3)^2 x = 0$. Thus

$$x = c_1 e^{-3t} + c_2 t e^{-3t}$$

$$y = -(2c_1 + c_2)e^{-3t} - 2c_2 t e^{-3t}.$$

Using $x(1) = 0$ and $y(1) = 1$ we obtain

$$c_1 e^{-3} + c_2 e^{-3} = 0$$

$$-(2c_1 + c_2)e^{-3} - 2c_2 e^{-3} = 1$$

or

$$c_1 + c_2 = 0$$

$$2c_1 + 3c_2 = -e^3.$$

Thus $c_1 = e^3$ and $c_2 = -e^3$. The solution of the initial value problem is

$$x = e^{-3t+3} - t e^{-3t+3}$$

$$y = -e^{-3t+3} + 2t e^{-3t+3}.$$

27. (*Suggestion*) Start by solving the first DE for x_1 and substituting the solution into the second DE.

Section 4.9

Nonlinear Differential Equations

The terminology and concepts listed below provide an outline of the main ideas encountered in this section. These can be useful when preparing for a quiz or test.

Terminology and Concepts

- superposition of solutions does *not* apply to nonlinear DEs

- nonlinear DEs can possess singular solutions

- using a substitution, express a nonlinear second-order DE, $F(x, y', y'') = 0$, as a first-order DE

- using a substitution, express a nonlinear second-order DE, $F(y, y', y'') = 0$, as a first-order DE

- Taylor series solution of an IVP

The basic skills listed below summarize the more mechanical types of problems encountered in the exercise set for this section.

Basic Skills

- solve a nonlinear second-order DE using reduction of order when appropriate
- use a substitution to find a Taylor series solution of a nonlinear second-order IVP

Exercises 4.9   *Hints, Suggestions, Solutions, and Examples*

3. Let $u = y'$ so that $u' = y''$. The equation becomes $u' = -u^2 - 1$ which is separable. Thus

$$\frac{du}{u^2 + 1} = -dx \implies \tan^{-1} u = -x + c_1 \implies y' = \tan(c_1 - x) \implies y = \ln|\cos(c_1 - x)| + c_2.$$

6. Let $u = y'$ so that $y'' = u\, du/dy$. The equation becomes $(y + 1)u\, du/dy = u^2$. Separating variables we obtain

$$\frac{du}{u} = \frac{dy}{y + 1} \implies \ln|u| = \ln|y + 1| + \ln c_1 \implies u = c_1(y + 1)$$

$$\implies \frac{dy}{dx} = c_1(y + 1) \implies \frac{dy}{y + 1} = c_1\, dx$$

$$\implies \ln|y + 1| = c_1 x + c_2 \implies y + 1 = c_3 e^{c_1 x}.$$

9. (a)

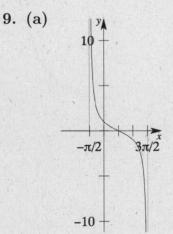

(b) Let $u = y'$ so that $y'' = u\, du/dy$. The equation becomes $u\, du/dy + yu = 0$. Separating variables we obtain

$$du = -y\, dy \implies u = -\frac{1}{2}y^2 + c_1 \implies y' = -\frac{1}{2}y^2 + c_1.$$

When $x = 0$, $y = 1$ and $y' = -1$ so $-1 = -1/2 + c_1$ and $c_1 = -1/2$. Then

$$\frac{dy}{dx} = -\frac{1}{2}y^2 - \frac{1}{2} \implies \frac{dy}{y^2 + 1} = -\frac{1}{2}\, dx \implies \tan^{-1} y = -\frac{1}{2}x + c_2$$

$$\implies y = \tan\left(-\frac{1}{2}x + c_2\right).$$

When $x = 0$, $y = 1$ so $1 = \tan c_2$ and $c_2 = \pi/4$. The solution of the initial-value problem is

$$y = \tan\left(\frac{\pi}{4} - \frac{1}{2}x\right).$$

The graph is shown in part (a).

(c) The interval of definition is $-\pi/2 < \pi/4 - x/2 < \pi/2$ or $-\pi/2 < x < 3\pi/2$.

12. Let $u = y'$ so that $u' = y''$. The equation becomes $u' - (1/x)u = u^2$, which is a Bernoulli DE. Using the substitution $w = u^{-1}$ we obtain $dw/dx + (1/x)w = -1$. An integrating factor is x, so

$$\frac{d}{dx}[xw] = -x \implies w = -\frac{1}{2}x + \frac{1}{x}c \implies \frac{1}{u} = \frac{c_1 - x^2}{2x} \implies u = \frac{2x}{c_1 - x^2} \implies y = -\ln\left|c_1 - x^2\right| + c_2.$$

15. We look for a solution of the form

$$y(x) = y(0) + y'(0)x + \frac{1}{2!}y''(0)x^2 + \frac{1}{3!}y'''(0)x^3 + \frac{1}{4!}y^{(4)}(0)x^4 + \frac{1}{5!}y^{(5)}(0)x^5.$$

From $y''(x) = x^2 + y^2 - 2y'$ we compute

$$y'''(x) = 2x + 2yy' - 2y''$$

$$y^{(4)}(x) = 2 + 2(y')^2 + 2yy'' - 2y'''$$

$$y^{(5)}(x) = 6y'y'' + 2yy''' - 2y^{(4)}.$$

Using $y(0) = 1$ and $y'(0) = 1$ we find

$$y''(0) = -1, \quad y'''(0) = 4, \quad y^{(4)}(0) = -6, \quad y^{(5)}(0) = 14.$$

An approximate solution is

$$y(x) = 1 + x - \frac{1}{2}x^2 + \frac{2}{3}x^3 - \frac{1}{4}x^4 + \frac{7}{60}x^5.$$

In the graph, the thinner curve is obtained using a numerical solver, while the thicker curve is the graph of the Taylor polynomial.

Chapter 4 in Review

Hints, Suggestions, Solutions, and Examples

3. It is not true unless the DE is homogeneous. For example, $y_1 = x$ is a solution of $y'' + y = x$, but $y_2 = 5x$ is not.

6. (a) Since $f_2(x) = 2\ln x = 2f_1(x)$, the set of functions is linearly dependent.

(b) Since x^{n+1} is not a constant multiple of x^n, the set of functions is linearly independent.

(c) Since $x + 1$ is not a constant multiple of x, the set of functions is linearly independent.

(d) Since $f_1(x) = \cos x \cos(\pi/2) - \sin x \sin(\pi/2) = -\sin x = -f_2(x)$, the set of functions is linearly dependent.

(e) Since $f_1(x) = 0 \cdot f_2(x)$, the set of functions is linearly dependent.

(f) Since $2x$ is not a constant multiple of 2, the set of functions is linearly independent.

(g) Since $3(x^2) + 2(1 - x^2) - (2 + x^2) = 0$, the set of functions is linearly dependent.

(h) Since $xe^{x+1} + 0(4x - 5)e^x - exe^x = 0$, the set of functions is linearly dependent.

9. From $m^2 - 2m - 2 = 0$ we obtain $m = 1 \pm \sqrt{3}$ so that

$$y = c_1 e^{(1+\sqrt{3})x} + c_2 e^{(1-\sqrt{3})x}.$$

12. From $2m^3 + 9m^2 + 12m + 5 = 0$ we obtain $m = -1$, $m = -1$, and $m = -5/2$ so that

$$y = c_1 e^{-5x/2} + c_2 e^{-x} + c_3 x e^{-x}.$$

15. Applying D^4 to the DE we obtain $D^4(D^2 - 3D + 5) = 0$. Then

$$y = \underbrace{e^{3x/2}\left(c_1 \cos \frac{\sqrt{11}}{2}x + c_2 \sin \frac{\sqrt{11}}{2}x\right)}_{y_c} + c_3 + c_4 x + c_5 x^2 + c_6 x^3$$

and $y_p = A + Bx + Cx^2 + Dx^3$. Substituting y_p into the DE yields

$$(5A - 3B + 2C) + (5B - 6C + 6D)x + (5C - 9D)x^2 + 5Dx^3 = -2x + 4x^3.$$

Equating coefficients gives $A = -222/625$, $B = 46/125$, $C = 36/25$, and $D = 4/5$. The general solution is

$$y = e^{3x/2}\left(c_1 \cos \frac{\sqrt{11}}{2}x + c_2 \sin \frac{\sqrt{11}}{2}x\right) - \frac{222}{625} + \frac{46}{125}x + \frac{36}{25}x^2 + \frac{4}{5}x^3.$$

18. Applying D to the DE we obtain $D(D^3 - D^2) = D^3(D - 1) = 0$. Then

$$y = \underbrace{c_1 + c_2 x + c_3 e^x}_{y_c} + c_4 x^2$$

and $y_p = Ax^2$. Substituting y_p into the DE yields $-2A = 6$. Equating coefficients gives $A = -3$. The general solution is

$$y = c_1 + c_2 x + c_3 e^x - 3x^2.$$

21. The auxiliary equation is $6m^2 - m - 1 = 0$ so that

$$y = c_1 x^{1/2} + c_2 x^{-1/3}.$$

24. The auxiliary equation is $m^2 - 2m + 1 = (m-1)^2 = 0$ and a particular solution is $y_p = \frac{1}{4}x^3$ so that

$$y = c_1 x + c_2 x \ln x + \frac{1}{4}x^3.$$

27. (a) The auxiliary equation is $m^4 - 2m^2 + 1 = (m^2 - 1)^2 = 0$, so the general solution of the DE is

$$y = c_1 \sinh x + c_2 \cosh x + c_3 x \sinh x + c_4 x \cosh x.$$

(b) Since both $\sinh x$ and $x \sinh x$ are solutions of the associated homogeneous DE, a particular solution of $y^{(4)} - 2y'' + y = \sinh x$ has the form $y_p = Ax^2 \sinh x + Bx^2 \cosh x$.

30. The auxiliary equation is $m^2 + 2m + 1 = (m+1)^2 = 0$, so that $y = c_1 e^{-x} + c_2 x e^{-x}$. Setting $y(-1) = 0$ and $y'(0) = 0$ we get $c_1 e - c_2 e = 0$ and $-c_1 + c_2 = 0$. Thus $c_1 = c_2$ and $y = c_1(e^{-x} + xe^{-x})$ is a solution of the boundary-value problem for any real number c_1.

33. Let $u = y'$ so that $u' = y''$. The equation becomes $u \, du/dx = 4x$. Separating variables we obtain

$$u \, du = 4x \, dx \implies \frac{1}{2}u^2 = 2x^2 + c_1 \implies u^2 = 4x^2 + c_2.$$

When $x = 1$, $y' = u = 2$, so $4 = 4 + c_2$ and $c_2 = 0$. Then

$$u^2 = 4x^2 \implies \frac{dy}{dx} = 2x \quad \text{or} \quad \frac{dy}{dx} = -2x$$

$$\implies y = x^2 + c_3 \quad \text{or} \quad y = -x^2 + c_4.$$

When $x = 1$, $y = 5$, so $5 = 1 + c_3$ and $5 = -1 + c_4$. Thus $c_3 = 4$ and $c_4 = 6$. We have $y = x^2 + 4$ and $y = -x^2 + 6$. Note however that when $y = -x^2 + 6$, $y' = -2x$ and $y'(1) = -2 \neq 2$. Thus, the solution of the initial-value problem is $y = x^2 + 4$.

36. Consider $xy'' + y' = 0$ and look for a solution of the form $y = x^m$. Substituting into the DE we have

$$xy'' + y' = m(m-1)x^{m-1} + mx^{m-1} = m^2 x^{m-1}.$$

Thus, the general solution of $xy'' + y' = 0$ is $y_c = c_1 + c_2 \ln x$. To find a particular solution of $xy'' + y' = -\sqrt{x}$ we use variation of parameters. The Wronskian is

$$W = \begin{vmatrix} 1 & \ln x \\ 0 & 1/x \end{vmatrix} = \frac{1}{x}.$$

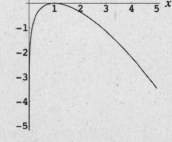

Identifying $f(x) = -x^{-1/2}$ we obtain

$$u_1' = \frac{x^{-1/2}\ln x}{1/x} = \sqrt{x}\ln x \quad \text{and} \quad u_2' = \frac{-x^{-1/2}}{1/x} = -\sqrt{x},$$

so that

$$u_1 = x^{3/2}\left(\frac{2}{3}\ln x - \frac{4}{9}\right) \quad \text{and} \quad u_2 = -\frac{2}{3}x^{3/2}.$$

Then

$$y_p = x^{3/2}\left(\frac{2}{3}\ln x - \frac{4}{9}\right) - \frac{2}{3}x^{3/2}\ln x = -\frac{4}{9}x^{3/2}$$

and the general solution of the DE is

$$y = c_1 + c_2\ln x - \frac{4}{9}x^{3/2}.$$

The initial conditions are $y(1) = 0$ and $y'(1) = 0$. These imply that $c_1 = \frac{4}{9}$ and $c_2 = \frac{2}{3}$. The solution of the IVP is

$$y = \frac{4}{9} + \frac{2}{3}\ln x - \frac{4}{9}x^{3/2}.$$

The graph is shown above.

39. From $(D-2)x - y = -e^t$ and $-3x + (D-4)y = -7e^t$ we obtain $(D-1)(D-5)x = -4e^t$ so that

$$x = c_1 e^t + c_2 e^{5t} + te^t.$$

Then

$$y = (D-2)x + e^t = -c_1 e^t + 3c_2 e^{5t} - te^t + 2e^t.$$

5 Modeling with Higher-Order Differential Equations

Linear Models: Initial-Value Problems

The terminology and concepts listed below provide an outline of the main ideas encountered in this section. These can be useful when preparing for a quiz or test.

Terminology and Concepts

- Hooke's law

- spring constant (k)

- Newton's second law of motion

- mass (m) versus weight $(W = mg)$

- equilibrium position

- DE of free undamped motion: $m\, d^2x/dt^2 + kx = 0$ or $d^2x/dt^2 + \omega^2 x = 0$, where $\omega^2 = k/m$

- equation of free undamped (simple harmonic) motion: $x(t) = c_1 \cos \omega t + c_2 \sin \omega t$

- period of motion: $T = 2\pi/\omega$

- cycle

- frequency of motion: $f = \omega/2\pi$

- amplitude of free vibrations (A)

- phase angle (ϕ)

- damping constant (β)

- DE of free damped motion: $m\, d^2x/dt^2 + \beta\, dx/dt + kx = 0$ or $d^2x/dt^2 + 2\lambda\, dx/dt + \omega^2 x = 0$ where $2\lambda = \beta/m$ and $\omega^2 = k/m$

- equation of motion of an overdamped system: $x(t) = e^{-\lambda t}(c_1 e^{\sqrt{\lambda^2 - \omega^2}\, t} + c_2 e^{-\sqrt{\lambda^2 - \omega^2}\, t})$

- equation of motion of a critically damped system: $x(t) = e^{-\lambda t}(c_1 + c_2 t)$

106

- equation of motion of an underdamped system:

$$x(t) = e^{-\lambda t}(c_1 \cos \sqrt{\omega^2 - \lambda^2}\, t + c_2 \sin \sqrt{\omega^2 - \lambda^2}\, t)$$

- damped amplitude: $Ae^{-\lambda t}$
- quasi period: $2\pi / \sqrt{\omega^2 - \lambda^2}$
- DE of driven (forced) motion with damping: $m\, d^2x/dt^2 + \beta\, dx/dt + kx = f(t)$
- transient term (solution)
- steady-state term (solution)
- DE of driven motion without damping: $m\, d^2x/dt^2 + kx = f(t)$
- pure resonance
- aging spring
- Airy's differential equation
- LRC series electrical circuit
- free electrical vibrations
- reactance
- impedance

The basic skills listed below summarize the more mechanical types of problems encountered in the exercise set for this section.

Basic Skills

- set up and solve DEs modeling free undamped motion of a spring/mass system
- set up and solve DEs modeling free damped motion of a spring/mass system
- set up and solve DEs modeling driven (forced) motion of a spring/mass system
- set up and solve DEs modeling an LRC series circuit

Phase Angle As pointed out in the text, be careful in computing the phase angle ϕ in a solution of the form $x(t) = A\sin(\omega t + \phi)$. The problem with using the inverse tangent $\tan^{-1} X$ is that all calculators are programmed for this particular inverse trigonometric function to give angular values that are either in the first or fourth quadrants. In other words, $-\pi/2 < \tan^{-1} X < \pi/2$. For example, because of periodicity there are infinitely many angles for which $\tan\phi = 1$ and for which $\tan\phi = -1/\sqrt{3}$. But there is only *one* answer when we use the notation $\tan^{-1} 1$ and $\tan^{-1}(-1/\sqrt{3})$. The exact values are:

$$\tan^{-1} 1 = \frac{\pi}{4} \quad \text{(first quadrant) because } \tan\frac{\pi}{4} = 1,$$

$$\tan^{-1}\left(-\frac{1}{\sqrt{3}}\right) = -\frac{\pi}{6} \quad \text{(fourth quadrant) because } \tan\left(-\frac{\pi}{6}\right) = -\frac{1}{\sqrt{3}}.$$

Of course, if you use a calculator you will either get

$$\tan^{-1} 1 = 45° \qquad \text{or} \qquad \tan^{-1} 1 = 0.78539816\ldots$$

depending on whether your calculator is set in degree or radian mode. The decimal number $0.78539816\ldots$ is the number π divided by 4. Now in order to compute ϕ in $x(t) = A\sin(\omega t + \phi)$ we must take into consideration all three components given in (7) of Section 5.1 in the text. Since $A > 0$, the signs of $\sin\phi = c_1/A$ and $\cos\phi = c_2/A$ are determined from the algebraic signs of c_1 and c_2 and these signs, in turn, indicate the quadrant in which ϕ lies. We summarize the four possibilities:

$$\phi = \tan^{-1}\left(\frac{c_1}{c_2}\right) \quad \text{if} \quad \overbrace{c_1 > 0,\ c_2 > 0}^{\phi \text{ in 1st quadrant}} \quad \text{or if} \quad \overbrace{c_1 < 0,\ c_2 > 0}^{\phi \text{ in 4th quadrant}}$$

$$\phi = \pi + \tan^{-1}\left(\frac{c_1}{c_2}\right) \quad \text{if} \quad \overbrace{c_1 > 0,\ c_2 < 0}^{\phi \text{ in 2nd quadrant}} \quad \text{or if} \quad \overbrace{c_1 < 0,\ c_2 < 0}^{\phi \text{ in 3rd quadrant}}.$$

Damping   In Examples 4 and 5 in the text, the statements

"...a damping force numerically equal to 2 times the instantaneous velocity..."

and

"...surrounding medium offers a resistance numerically equal to the instantaneous velocity..."

are to be interpreted: In Example 4, the damping constant is $\beta = 2$ in equation (10) and in Example 5, the damping constant is $\beta = 1$ in equation (10).

Exercises 5.1   *Hints, Suggestions, Solutions, and Examples*

3. From $\frac{3}{4}x'' + 72x = 0$, $x(0) = -1/4$, and $x'(0) = 0$ we obtain $x = -\frac{1}{4}\cos 4\sqrt{6}\,t$.

6. From $50x'' + 200x = 0$, $x(0) = 0$, and $x'(0) = -10$ we obtain $x = -5\sin 2t$ and $x' = -10\cos 2t$.

9. From $\frac{1}{4}x'' + x = 0$, $x(0) = 1/2$, and $x'(0) = 3/2$ we obtain

$$x = \frac{1}{2}\cos 2t + \frac{3}{4}\sin 2t = \frac{\sqrt{13}}{4}\sin(2t + 0.588).$$

10-11. (*Suggestion*) Read the material above in this section on **Phase Angle**.

12. From $x'' + 9x = 0$, $x(0) = -1$, and $x'(0) = -\sqrt{3}$ we obtain

$$x = -\cos 3t - \frac{\sqrt{3}}{3}\sin 3t = \frac{2}{\sqrt{3}}\sin\left(3t + \frac{4\pi}{3}\right).$$

and $x' = 2\sqrt{3}\cos(3t + 4\pi/3)$. If $x' = 3$ then $t = -7\pi/18 + 2n\pi/3$ and $t = -\pi/2 + 2n\pi/3$ for $n = 1, 2, 3, \ldots$.

15. For large values of t the DE is approximated by $x'' = 0$. The solution of this equation is the linear function $x = c_1 t + c_2$. Thus, for large time, the restoring force will have decayed to the point where the spring is incapable of returning the mass, and the spring will simply keep on stretching.

18. (a) below   **(b)** from rest

21. From $\frac{1}{8}x'' + x' + 2x = 0$, $x(0) = -1$, and $x'(0) = 8$ we obtain $x = 4te^{-4t} - e^{-4t}$ and $x' = 8e^{-4t} - 16te^{-4t}$. If $x = 0$ then $t = 1/4$ second. If $x' = 0$ then $t = 1/2$ second and the extreme displacement is $x = e^{-2}$ feet.

24. (a) $x = \frac{1}{3}e^{-8t}\left(4e^{6t} - 1\right)$ is not zero for $t \geq 0$; the extreme displacement is $x(0) = 1$ meter.

(b) $x = \frac{1}{3}e^{-8t}\left(5 - 2e^{6t}\right) = 0$ when $t = \frac{1}{6}\ln\frac{5}{2} \approx 0.153$ second; if $x' = \frac{4}{3}e^{-8t}\left(e^{6t} - 10\right) = 0$ then $t = \frac{1}{6}\ln 10 \approx 0.384$ second and the extreme displacement is $x = -0.232$ meter.

27. From $\frac{5}{16}x'' + \beta x' + 5x = 0$ we find that the roots of the auxiliary equation are $m = -\frac{8}{5}\beta \pm \frac{4}{5}\sqrt{4\beta^2 - 25}$.

(a) If $4\beta^2 - 25 > 0$ then $\beta > 5/2$.

(b) If $4\beta^2 - 25 = 0$ then $\beta = 5/2$.

(c) If $4\beta^2 - 25 < 0$ then $0 < \beta < 5/2$.

30. (a) If $x'' + 2x' + 5x = 12\cos 2t + 3\sin 2t$, $x(0) = 1$, and $x'(0) = 5$ then $x_c = e^{-t}(c_1\cos 2t + c_2\sin 2t)$ and $x_p = 3\sin 2t$ so that the equation of motion is

$$x = e^{-t}\cos 2t + 3\sin 2t.$$

(b)   **(c)**

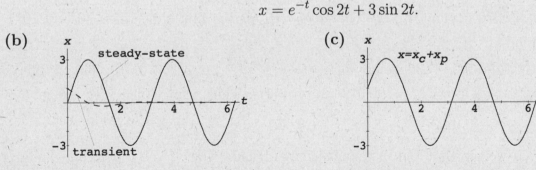

33. From $2x'' + 32x = 68e^{-2t}\cos 4t$, $x(0) = 0$, and $x'(0) = 0$ we obtain $x_c = c_1\cos 4t + c_2\sin 4t$ and $x_p = \frac{1}{2}e^{-2t}\cos 4t - 2e^{-2t}\sin 4t$ so that

$$x = -\frac{1}{2}\cos 4t + \frac{9}{4}\sin 4t + \frac{1}{2}e^{-2t}\cos 4t - 2e^{-2t}\sin 4t.$$

36. (a) From $100x'' + 1600x = 1600\sin 8t$, $x(0) = 0$, and $x'(0) = 0$ we obtain $x_c = c_1\cos 4t + c_2\sin 4t$ and $x_p = -\frac{1}{3}\sin 8t$ so that by a trig identity

$$x = \frac{2}{3}\sin 4t - \frac{1}{3}\sin 8t = \frac{2}{3}\sin 4t - \frac{2}{3}\sin 4t\cos 4t.$$

109

(b) If $x = \frac{1}{3}\sin 4t(2 - 2\cos 4t) = 0$ then $t = n\pi/4$ for $n = 0, 1, 2, \ldots$.

(c) If $x' = \frac{8}{3}\cos 4t - \frac{8}{3}\cos 8t = \frac{8}{3}(1 - \cos 4t)(1 + 2\cos 4t) = 0$ then $t = \pi/3 + n\pi/2$ and $t = \pi/6 + n\pi/2$ for $n = 0, 1, 2, \ldots$ at the extreme values. *Note*: There are many other values of t for which $x' = 0$.

(d) $x(\pi/6 + n\pi/2) = \sqrt{3}/2$ cm and $x(\pi/3 + n\pi/2) = -\sqrt{3}/2$ cm

(e)

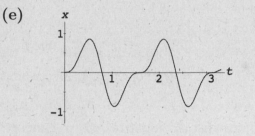

39. (a) From $x'' + \omega^2 x = F_0\cos\gamma t$, $x(0) = 0$, and $x'(0) = 0$ we obtain $x_c = c_1\cos\omega t + c_2\sin\omega t$ and $x_p = (F_0\cos\gamma t)/\left(\omega^2 - \gamma^2\right)$ so that

$$x = -\frac{F_0}{\omega^2 - \gamma^2}\cos\omega t + \frac{F_0}{\omega^2 - \gamma^2}\cos\gamma t.$$

(b) $\displaystyle\lim_{\gamma\to\omega}\frac{F_0}{\omega^2 - \gamma^2}(\cos\gamma t - \cos\omega t) = \lim_{\gamma\to\omega}\frac{-F_0 t\sin\gamma t}{-2\gamma} = \frac{F_0}{2\omega}t\sin\omega t.$

45. Solving $\frac{1}{20}q'' + 2q' + 100q = 0$ we obtain $q(t) = e^{-20t}(c_1\cos 40t + c_2\sin 40t)$. The initial conditions $q(0) = 5$ and $q'(0) = 0$ imply $c_1 = 5$ and $c_2 = 5/2$. Thus

$$q(t) = e^{-20t}\left(5\cos 40t + \frac{5}{2}\sin 40t\right) = \sqrt{25 + 25/4}\,e^{-20t}\sin(40t + 1.1071)$$

and $q(0.01) \approx 4.5676$ coulombs. The charge is zero for the first time when $40t + 1.1071 = \pi$ or $t \approx 0.0509$ second.

48. Solving $q'' + 100q' + 2500q = 30$ we obtain $q(t) = c_1 e^{-50t} + c_2 te^{-50t} + 0.012$. The initial conditions $q(0) = 0$ and $q'(0) = 2$ imply $c_1 = -0.012$ and $c_2 = 1.4$. Thus, using $i(t) = q'(t)$ we get

$$q(t) = -0.012e^{-50t} + 1.4te^{-50t} + 0.012 \quad\text{and}\quad i(t) = 2e^{-50t} - 70te^{-50t}.$$

Solving $i(t) = 0$ we see that the maximum charge occurs when $t = 1/35$ second and $q(1/35) \approx 0.01871$ coulomb.

51. The DE is $\frac{1}{2}q'' + 20q' + 1000q = 100\sin 60t$. To use Example 10 in the text we identify $E_0 = 100$ and $\gamma = 60$. Then

$$X = L\gamma - \frac{1}{c\gamma} = \frac{1}{2}(60) - \frac{1}{0.001(60)} \approx 13.3333,$$

$$Z = \sqrt{X^2 + R^2} = \sqrt{X^2 + 400} \approx 24.0370,$$

and

$$\frac{E_0}{Z} = \frac{100}{Z} \approx 4.1603.$$

From Problem 50, then

$$i_p(t) \approx 4.1603 \sin(60t + \phi)$$

where $\sin \phi = -X/Z$ and $\cos \phi = R/Z$. Thus $\tan \phi = -X/R \approx -0.6667$ and ϕ is a fourth quadrant angle. Now $\phi \approx -0.5880$ and

$$i_p(t) = 4.1603 \sin(60t - 0.5880).$$

54. In Problem 50 it is shown that the amplitude of the steady-state current is E_0/Z, where $Z = \sqrt{X^2 + R^2}$ and $X = L\gamma - 1/C\gamma$. Since E_0 is constant the amplitude will be a maximum when Z is a minimum. Since R is constant, Z will be a minimum when $X = 0$. Solving $L\gamma - 1/C\gamma = 0$ for γ we obtain $\gamma = 1/\sqrt{LC}$. The maximum amplitude will be E_0/R.

57. In an LC-series circuit there is no resistor, so the DE is

$$L\frac{d^2q}{dt^2} + \frac{1}{C}q = E(t).$$

Then $q(t) = c_1 \cos\left(t/\sqrt{LC}\right) + c_2 \sin\left(t/\sqrt{LC}\right) + q_p(t)$ where $q_p(t) = A\sin\gamma t + B\cos\gamma t$. Substituting $q_p(t)$ into the DE we find

$$\left(\frac{1}{C} - L\gamma^2\right) A\sin\gamma t + \left(\frac{1}{C} - L\gamma^2\right) B\cos\gamma t = E_0 \cos\gamma t.$$

Equating coefficients we obtain $A = 0$ and $B = E_0 C/(1 - LC\gamma^2)$. Thus, the charge is

$$q(t) = c_1 \cos\frac{1}{\sqrt{LC}}t + c_2 \sin\frac{1}{\sqrt{LC}}t + \frac{E_0 C}{1 - LC\gamma^2}\cos\gamma t.$$

The initial conditions $q(0) = q_0$ and $q'(0) = i_0$ imply $c_1 = q_0 - E_0C/(1 - LC\gamma^2)$ and $c_2 = i_0\sqrt{LC}$. The current is $i(t) = q'(t)$ or

$$i(t) = -\frac{c_1}{\sqrt{LC}}\sin\frac{1}{\sqrt{LC}}t + \frac{c_2}{\sqrt{LC}}\cos\frac{1}{\sqrt{LC}}t - \frac{E_0 C\gamma}{1 - LC\gamma^2}\sin\gamma t$$

$$= i_0 \cos\frac{1}{\sqrt{LC}}t - \frac{1}{\sqrt{LC}}\left(q_0 - \frac{E_0 C}{1 - LC\gamma^2}\right)\sin\frac{1}{\sqrt{LC}}t - \frac{E_0 C\gamma}{1 - LC\gamma^2}\sin\gamma t.$$

Section 5.2

Linear Models: Boundary-Value Problems

The terminology and concepts listed below provide an outline of the main ideas encountered in this section. These can be useful when preparing for a quiz or test.

Terminology and Concepts

- axis of symmetry of a beam
- deflection curve or elastic curve of a beam
- cantilever beam
- eigenvalues and corresponding eigenfunctions of a boundary-value problem

The basic skills listed below summarize the more mechanical types of problems encountered in the exercise set for this section.

Basic Skills

- solve problems involving beams satisfying various end (boundary) conditions such as embedded, free, or simply supported
- find the eigenvalues and corresponding eigenfunctions for second-order boundary-value problems

Exercises 5.2   *Hints, Suggestions, Solutions, and Examples*

3. (a) The general solution is

$$y(x) = c_1 + c_2 x + c_3 x^2 + c_4 x^3 + \frac{w_0}{24EI} x^4.$$

The boundary conditions are $y(0) = 0$, $y'(0) = 0$, $y(L) = 0$, $y''(L) = 0$. The first two conditions give $c_1 = 0$ and $c_2 = 0$. The conditions at $x = L$ give the system

$$c_3 L^2 + c_4 L^3 + \frac{w_0}{24EI} L^4 = 0$$

$$2c_3 + 6c_4 L + \frac{w_0}{2EI} L^2 = 0.$$

Solving, we obtain $c_3 = w_0 L^2/16EI$ and $c_4 = -5w_0 L/48EI$. The deflection is

$$y(x) = \frac{w_0}{48EI}(3L^2 x^2 - 5Lx^3 + 2x^4).$$

(b)

6. **(a)** $y_{max} = y(L) = w_0 L^4 / 8EI$

   (b) Replacing both L and x by $L/2$ in $y(x)$ we obtain $w_0 L^4 / 128EI$, which is 1/16 of the maximum deflection when the length of the beam is L.

   (c) $y_{max} = y(L/2) = 5w_0 L^4 / 384EI$

   (d) The maximum deflection in Example 1 is $y(L/2) = (w_0/24EI)L^4/16 = w_0 L^4/384EI$, which is 1/5 of the maximum displacement of the beam in part (c).

9. This is Example 2 in the text with $L = \pi$. The eigenvalues are $\lambda_n = n^2\pi^2/\pi^2 = n^2$, $n = 1, 2, 3, \ldots$ and the corresponding eigenfunctions are $y_n = \sin(n\pi x/\pi) = \sin nx$, $n = 1, 2, 3, \ldots$.

12. For $\lambda \leq 0$ the only solution of the boundary-value problem is $y = 0$. For $\lambda = \alpha^2 > 0$ we have

$$y = c_1 \cos \alpha x + c_2 \sin \alpha x.$$

Since $y(0) = 0$ implies $c_1 = 0$, $y = c_2 \sin x\, dx$. Now

$$y'\left(\frac{\pi}{2}\right) = c_2 \alpha \cos \alpha \frac{\pi}{2} = 0$$

gives

$$\alpha \frac{\pi}{2} = \frac{(2n-1)\pi}{2} \quad \text{or} \quad \lambda = \alpha^2 = (2n-1)^2, \ n = 1, 2, 3, \ldots.$$

The eigenvalues $\lambda_n = (2n-1)^2$ correspond to the eigenfunctions $y_n = \sin(2n-1)x$.

15. The auxiliary equation has solutions

$$m = \frac{1}{2}\left(-2 \pm \sqrt{4 - 4(\lambda + 1)}\right) = -1 \pm \alpha.$$

For $\lambda = -\alpha^2 < 0$ we have

$$y = e^{-x}\left(c_1 \cosh \alpha x + c_2 \sinh \alpha x\right).$$

The boundary conditions imply

$$y(0) = c_1 = 0$$

$$y(5) = c_2 e^{-5} \sinh 5\alpha = 0$$

so $c_1 = c_2 = 0$ and the only solution of the BVP is $y = 0$.

113

For $\lambda = 0$ we have

$$y = c_1 e^{-x} + c_2 x e^{-x}$$

and the only solution of the BVP is $y = 0$.

For $\lambda = \alpha^2 > 0$ we have

$$y = e^{-x} \left(c_1 \cos \alpha x + c_2 \sin \alpha x \right).$$

Now $y(0) = 0$ implies $c_1 = 0$, so

$$y(5) = c_2 e^{-5} \sin 5\alpha = 0$$

gives

$$5\alpha = n\pi \quad \text{or} \quad \lambda = \alpha^2 = \frac{n^2 \pi^2}{25}, \quad n = 1, 2, 3, \ldots.$$

The eigenvalues $\lambda_n = \dfrac{n^2 \pi^2}{25}$ correspond to the eigenfunctions $y_n = e^{-x} \sin \dfrac{n\pi}{5} x$ for $n = 1, 2, 3, \ldots$.

16. (*Suggestion*) Let $\lambda + 1 = 0$, $\lambda + 1 = -\alpha^2$, and $\lambda + 1 = \alpha^2$.

18. For $\lambda = 0$ the general solution is $y = c_1 + c_2 \ln x$. Now $y' = c_2/x$, so $y'(e^{-1}) = c_2 e = 0$ implies $c_2 = 0$. Then $y = c_1$ and $y(1) = 0$ gives $c_1 = 0$. Thus $y(x) = 0$.

For $\lambda = -\alpha^2 < 0$, $y = c_1 x^{-\alpha} + c_2 x^{\alpha}$. The boundary conditions give $c_2 = c_1 e^{2\alpha}$ and $c_1 = 0$, so that $c_2 = 0$ and $y(x) = 0$.

For $\lambda = \alpha^2 > 0$, $y = c_1 \cos(\alpha \ln x) + c_2 \sin(\alpha \ln x)$. From $y(1) = 0$ we obtain $c_1 = 0$ and $y = c_2 \sin(\alpha \ln x)$. Now $y' = c_2 (\alpha/x) \cos(\alpha \ln x)$, so $y'(e^{-1}) = c_2 e \alpha \cos \alpha = 0$ implies $\cos \alpha = 0$ or $\alpha = (2n-1)\pi/2$ and $\lambda = \alpha^2 = (2n-1)^2 \pi^2 / 4$ for $n = 1, 2, 3, \ldots$. The corresponding eigenfunctions are

$$y_n = \sin \left(\frac{2n-1}{2} \pi \ln x \right).$$

21. If restraints are put on the column at $x = L/4$, $x = L/2$, and $x = 3L/4$, then the critical load will be P_4.

24. (a) The boundary-value problem is

$$\frac{d^4 y}{dx^4} + \lambda \frac{d^2 y}{dx^2} = 0, \quad y(0) = 0, y''(0) = 0, \; y(L) = 0, y'(L) = 0,$$

where $\lambda = \alpha^2 = P/EI$. The solution of the DE is $y = c_1 \cos \alpha x + c_2 \sin \alpha x + c_3 x + c_4$ and the conditions $y(0) = 0$, $y''(0) = 0$ yield $c_1 = 0$ and $c_4 = 0$. Next, by applying $y(L) = 0$, $y'(L) = 0$ to $y = c_2 \sin \alpha x + c_3 x$ we get the system of equations

$$c_2 \sin \alpha L + c_3 L = 0$$

$$\alpha c_2 \cos \alpha L + c_3 = 0.$$

To obtain nontrivial solutions c_2, c_3, we must have the determinant of the coefficients equal to zero:

$$\begin{vmatrix} \sin \alpha L & L \\ \alpha \cos \alpha L & 1 \end{vmatrix} = 0 \quad \text{or} \quad \tan \beta = \beta,$$

where $\beta = \alpha L$. If β_n denotes the positive roots of the last equation, then the eigenvalues are found from $\beta_n = \alpha_n L = \sqrt{\lambda_n} \, L$ or $\lambda_n = (\beta_n/L)^2$. From $\lambda = P/EI$ we see that the critical loads are $P_n = \beta_n^2 EI/L^2$. With the aid of a CAS we find that the first positive root of $\tan \beta = \beta$ is (approximately) $\beta_1 = 4.4934$, and so the Euler load is (approximately) $P_1 = 20.1907 EI/L^2$. Finally, if we use $c_3 = -c_2 \alpha \cos \alpha L$, then the deflection curves are

$$y_n(x) = c_2 \sin \alpha_n x + c_3 x = c_2 \left[\sin\left(\frac{\beta_n}{L} x\right) - \left(\frac{\beta_n}{L} \cos \beta_n\right) x \right].$$

(b) With $L = 1$ and c_2 appropriately chosen, the general shape of the first buckling mode,

$$y_1(x) = c_2 \left[\sin\left(\frac{4.4934}{L} x\right) - \left(\frac{4.4934}{L} \cos(4.4934)\right) x \right],$$

is shown below.

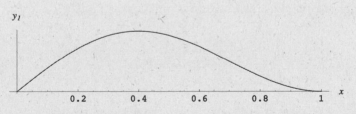

27. The auxiliary equation is $m^2 + m = m(m + 1) = 0$ so that $u(r) = c_1 r^{-1} + c_2$. The boundary conditions $u(a) = u_0$ and $u(b) = u_1$ yield the system $c_1 a^{-1} + c_2 = u_0$, $c_1 b^{-1} + c_2 = u_1$. Solving gives

$$c_1 = \left(\frac{u_0 - u_1}{b - a}\right) ab \quad \text{and} \quad c_2 = \frac{u_1 b - u_0 a}{b - a}.$$

Thus

$$u(r) = \left(\frac{u_0 - u_1}{b - a}\right) \frac{ab}{r} + \frac{u_1 b - u_0 a}{b - a}.$$

Section 5.3 — Nonlinear Models

Nonlinear Models

The terminology and concepts listed below provide an outline of the main ideas encountered in this section. These can be useful when preparing for a quiz or test.

Terminology and Concepts

- nonlinear spring models

- hard spring

- soft spring

- simple pendulum

- linearization of a DE

The basic skills listed below summarize the more mechanical types of problems encountered in the exercise set for this section.

Basic Skills

- use a numerical solver to plot solution curves of nonlinear DEs

Exercises 5.3   *Hints, Suggestions, Solutions, and Examples*

3. The period corresponding to $x(0) = 1$, $x'(0) = 1$ is approximately 5.8. The second initial-value problem does not have a periodic solution.

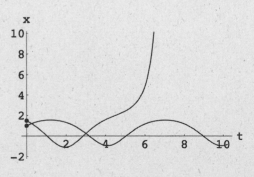

6. From the graphs we see that the interval is approximately $(-0.8, 1.1)$.

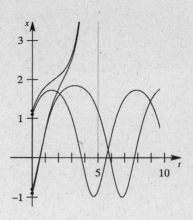

9. This is a damped hard spring, so x will approach 0 as t approaches ∞.

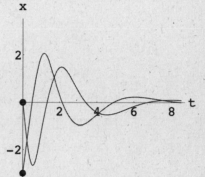

12. (a)

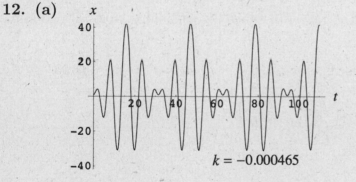

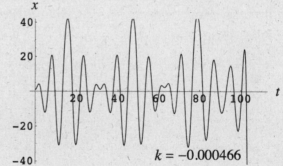

The system appears to be oscillatory for $-0.000465 \leq k_1 < 0$ and nonoscillatory for $k_1 \leq -0.000466$.

(b)

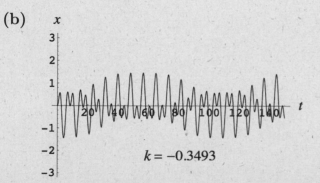

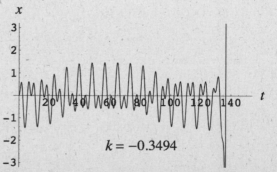

The system appears to be oscillatory for $-0.3493 \leq k_1 < 0$ and nonoscillatory for $k_1 \leq -0.3494$.

15. **(a)** Intuitively, one might expect that only half of a 10-pound chain could be lifted by a 5-pound vertical force.

    (b) Since $x = 0$ when $t = 0$, and $v = dx/dt = \sqrt{160 - 64x/3}$, we have $v(0) = \sqrt{160} \approx 12.65$ ft/s.

    (c) Since x should always be positive, we solve $x(t) = 0$, getting $t = 0$ and $t = \frac{3}{2}\sqrt{5/2} \approx 2.3717$. Since the graph of $x(t)$ is a parabola, the maximum value occurs at $t_m = \frac{3}{4}\sqrt{5/2}$. (This can also be obtained by solving $x'(t) = 0$.) At this time the height of the chain is $x(t_m) \approx 7.5$ ft. This is higher than predicted because of the momentum generated by the force. When the chain is 5 feet high it still has a positive velocity of about 7.3 ft/s, which keeps it going higher for a while.

    (d) As discussed in the solution to part (c) of this problem, the chain has momentum generated by the force applied to it that will cause it to go higher than expected. It will then fall back to below the expected maximum height, again due to momentum. This, in turn, will cause it to next go higher than expected, and so on.

18. **(a)** Let (r, θ) denote the polar coordinates of the destroyer S_1. When S_1 travels the 6 miles from $(9, 0)$ to $(3, 0)$ it stands to reason, since S_2 travels half as fast as S_1, that the polar coordinates of S_2 are $(3, \theta_2)$, where θ_2 is unknown. In other words, the distances of the ships from $(0, 0)$ are the same and $r(t) = 15t$ then gives the radial distance of both ships. This is necessary if S_1 is to intercept S_2.

    (b) The differential of arc length in polar coordinates is $(ds)^2 = (r\, d\theta)^2 + (dr)^2$, so that

    $$\left(\frac{ds}{dt}\right)^2 = r^2 \left(\frac{d\theta}{dt}\right)^2 + \left(\frac{dr}{dt}\right)^2.$$

    Using $ds/dt = 30$ and $dr/dt = 15$ then gives

    $$900 = 225t^2 \left(\frac{d\theta}{dt}\right)^2 + 225$$

    $$675 = 225t^2 \left(\frac{d\theta}{dt}\right)^2$$

    $$\frac{d\theta}{dt} = \frac{\sqrt{3}}{t}$$

    $$\theta(t) = \sqrt{3}\ln t + c = \sqrt{3}\ln\frac{r}{15} + c.$$

    When $r = 3$, $\theta = 0$, so $c = -\sqrt{3}\ln\frac{1}{5}$ and

    $$\theta(t) = \sqrt{3}\left(\ln\frac{r}{15} - \ln\frac{1}{5}\right) = \sqrt{3}\ln\frac{r}{3}.$$

Thus $r = 3e^{\theta/\sqrt{3}}$, whose graph is a logarithmic spiral.

(c) The time for S_1 to go from $(9,0)$ to $(3,0) = \frac{1}{5}$ hour. Now S_1 must intercept the path of S_2 for some angle β, where $0 < \beta < 2\pi$. At the time of interception t_2 we have $15t_2 = 3e^{\beta/\sqrt{3}}$ or $t = \frac{1}{5}e^{\beta/\sqrt{3}}$. The total time is then

$$t = \frac{1}{5} + \frac{1}{5}e^{\beta/\sqrt{3}} < \frac{1}{5}(1 + e^{2\pi/\sqrt{3}}).$$

Chapter 5 in Review

Hints, Suggestions, Solutions, and Examples

3. $5/4$ m, since $x = -\cos 4t + \frac{3}{4}\sin 4t$

6. False; since the equation of motion in this case is $x(t) = e^{-\lambda t}(c_1 + c_2 t)$ and $x(t) = 0$ can have at most one real solution

9. $y = 0$ because $\lambda = 8$ is not an eigenvalue

12. (a) Solving $\frac{3}{8}x'' + 6x = 0$ subject to $x(0) = 1$ and $x'(0) = -4$ we obtain

$$x = \cos 4t - \sin 4t = \sqrt{2}\sin\left(4t + 3\pi/4\right).$$

(b) The amplitude is $\sqrt{2}$, period is $\pi/2$, and frequency is $2/\pi$.

(c) If $x = 1$ then $t = n\pi/2$ and $t = -\pi/8 + n\pi/2$ for $n = 1, 2, 3, \ldots$.

(d) If $x = 0$ then $t = \pi/16 + n\pi/4$ for $n = 0, 1, 2, \ldots$. The motion is upward for n even and downward for n odd.

(e) $x'(3\pi/16) = 0$

(f) If $x' = 0$ then $4t + 3\pi/4 = \pi/2 + n\pi$ or $t = 3\pi/16 + n\pi$.

15. From $mx'' + 4x' + 2x = 0$ we see that nonoscillatory motion results if $16 - 8m \geq 0$ or $0 < m \leq 2$.

18. Clearly $x_p = A/\omega^2$ suffices.

21. From $q'' + 10^4 q = 100\sin 50t$, $q(0) = 0$, and $q'(0) = 0$ we obtain $q_c = c_1 \cos 100t + c_2 \sin 100t$, $q_p = \frac{1}{75}\sin 50t$, and

(a) $q = -\frac{1}{150}\sin 100t + \frac{1}{75}\sin 50t$,

(b) $i = -\frac{2}{3}\cos 100t + \frac{2}{3}\cos 50t$, and

(c) $q = 0$ when $\sin 50t(1 - \cos 50t) = 0$ or $t = n\pi/50$ for $n = 0, 1, 2, \ldots$.

24. (a) The DE is $d^2r/dt^2 - \omega^2 r = -g\sin\omega t$, and the auxiliary equation is $m^2 - \omega^2 = 0$, so $r_c = c_1 e^{\omega t} + c_2 e^{-\omega t}$. A particular solution has the form $r_p = A\sin\omega t + B\cos\omega t$. Substituting into the DE we find $-2A\omega^2\sin\omega t - 2B\omega^2\cos\omega t = -g\sin\omega t$. Thus, $B = 0$, $A = g/2\omega^2$, and $r_p = (g/2\omega^2)\sin\omega t$. The general solution of the DE is $r(t) = c_1 e^{\omega t} + c_2 e^{-\omega t} + (g/2\omega^2)\sin\omega t$. The initial conditions imply $c_1 + c_2 = r_0$ and $g/2\omega - \omega c_1 + \omega c_2 = v_0$. Solving for c_1 and c_2 we get

$$c_1 = (2\omega^2 r_0 + 2\omega v_0 - g)/4\omega^2 \quad \text{and} \quad c_2 = (2\omega^2 r_0 - 2\omega v_0 + g)/4\omega^2,$$

so that

$$r(t) = \frac{2\omega^2 r_0 + 2\omega v_0 - g}{4\omega^2} e^{\omega t} + \frac{2\omega^2 r_0 - 2\omega v_0 + g}{4\omega^2} e^{-\omega t} + \frac{g}{2\omega^2}\sin\omega t.$$

(b) The bead will exhibit simple harmonic motion when the exponential terms are missing. Solving $c_1 = 0$, $c_2 = 0$ for r_0 and v_0 we find $r_0 = 0$ and $v_0 = g/2\omega$.

To find the minimum length of rod that will accommodate simple harmonic motion we determine the amplitude of $r(t)$ and double it. Thus $L = g/\omega^2$.

(c) As t increases, $e^{\omega t}$ approaches infinity and $e^{-\omega t}$ approaches 0. Since $\sin\omega t$ is bounded, the distance, $r(t)$, of the bead from the pivot point increases without bound and the distance of the bead from P will eventually exceed $L/2$.

(d)

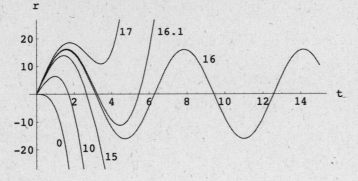

(e) For each v_0 we want to find the smallest value of t for which $r(t) = \pm 20$. Whether we look for $r(t) = -20$ or $r(t) = 20$ is determined by looking at the graphs in part (d). The total times that the bead stays on the rod is shown in the table below.

v_0	0	10	15	16.1	17
r	−20	−20	−20	20	20
t	1.55007	2.35494	3.43088	6.11627	4.22339

When $v_0 = 16$ the bead never leaves the rod.

6

Series Solutions of Linear Equations

Section 6.1

Solutions about Ordinary Points

The terminology and concepts listed below provide an outline of the main ideas encountered in this section. These can be useful when preparing for a quiz or test.

Terminology and Concepts

- power series centered at a; $\sum_{n=0}^{\infty} c_n(x-a)^n$
- convergence of a power series
- interval of convergence
- radius of convergence
- absolute convergence
- ratio test
- a power series defines a function
- a function analytic at a point
- ordinary point
- singular point
- power series solution of a DE
- recurrence relation

The basic skills listed below summarize the more mechanical types of problems encountered in the exercise set for this section.

Basic Skills

- find the radius of convergence of a power series
- convert an expression like $\sum_{n=3}^{\infty}(n+1)x^{n-2} + \sum_{n=2}^{\infty} nx^n$ to an expression involving a single infinite summation: $4x + \sum_{k=2}^{\infty}(2k+3)x^k$

- find power series solutions of linear second-order DEs

Summation Notation   Without a doubt, the key to successfully solving a linear DE with variable coefficients by the power series method depends on your ability to recast a combination of power series centered at the same point in terms of a *single* summation. In Example 1 of Section 6.1 we gave an example involving two series; here is another example of how this is done using three series. We wish to combine the three sums

$$\sum_{n=2}^{\infty} c_n n(n-1)x^{n-2} + 2\sum_{n=1}^{\infty} c_n n x^n + 5\sum_{n=0}^{\infty} c_n x^{n+1}$$

into one term. Notice that the first series starts (for $n=2$) with x^0, the second series starts (for $n=1$) with x^1, and the third series also starts (for $n=0$) with x^1. To bring the first series in line with the last two, we write out the term corresponding to $n=2$ and observe that now all three series

$$2c_2 + \sum_{n=3}^{\infty} c_n n(n-1)x^{n-2} + \sum_{n=1}^{\infty} 2c_n n x^n + \sum_{n=0}^{\infty} 5c_n x^{n+1}$$

begin with the same power x^1. The next thing that must be done is to get a common summation index for the series:

- In the first series, we let $k = n - 2$. As $n = 3, 4, 5, \ldots,$ k takes on the values $1, 2, 3, \ldots.$
- In the second series, we let $k = n$. Here $n = 1, 2, 3, \ldots$ corresponds to $k = 1, 2, 3, \ldots.$
- In the third series, we let $k = n + 1$. Then $n = 0, 1, 2, \ldots$ also yields $k = 1, 2, 3, \ldots.$

Next, we substitute $n = k + 2$ in the first series, $n = k$ in the second series, and then $n = k - 1$ in the third:

$$2c_2 + \sum_{k=1}^{\infty} c_{k+2}(k+2)(k+1)x^k + \sum_{k=1}^{\infty} 2c_k k x^k + \sum_{k=1}^{\infty} 5c_{k-1}x^k.$$

At this point, when all the series start with the same power of x and have the same summation index starting with the same value (in this case, $k = 1$), we can add the series term-by-term to get

$$2c_2 + \sum_{k=1}^{\infty} [c_{k+2}(k+2)(k+1) + 2c_k k + 5c_{k-1}]x^k.$$

Alternatively, we can make the change of summation variable first as illustrated in the solution of Problem 18 in this section of the manual.

Use of Computers   Both *Mathematica* and *Maple* can express the sum of several terms in a concise way without actually listing each term. The commands below can be used to express the first four terms of the infinite sum $\sum_{n=0}^{\infty} c_n (-1)^n/(n+1)^2.$

Clear[c] (*Mathematica*)

Sum[c[n](-1)^n/(n+1)^2, {n, 0, 3}]

```
sum(c(n)*(-1)^n/(n+1)^2, n=0..3);        (Maple)
```

To obtain the power series representation about $x = 0$ of a function like $f(x) = e^x$ use

Series[Exp[x], {x, 0, 5}] (*Mathematica*)

or

```
series(exp(x),x=0,5);                     (Maple)
```

The *Mathematica* output corresponding to the above command is

$$1 + x + \frac{1}{2}x^2 + \frac{1}{6}x^3 + \frac{1}{24}x^4 + \frac{1}{120}x^5 + O[x]^6,$$

whereas for *Maple* the output is

$$1 + x + \frac{1}{2}x^2 + \frac{1}{6}x^3 + \frac{1}{24}x^4 + O(x^5).$$

Note that in the *Mathematica* command the 5 tells the program to compute the series out to, and including, the term containing x^5. On the other hand, in *Maple*, the 5 tells the program to compute the series out to, but excluding, the term containing x^5.

Depending on how you want to use the series representation of a function found using a CAS, it may be inconvenient to have the order of the series included in the output. For example, neither *Mathematica* nor *Maple* will plot the graph of an expression containing $O[x]^6$, in the case of *Mathematica*, or $O(x^5)$ in the case of *Maple*. To get around this, both programs have commands that truncate the series, converting it to a polynomial.

Normal[Series[Exp[x], {x, 0, 5}]] (*Mathematica*)

or

Series[Exp[x], {x, 0, 5}]//Normal

```
convert(series(exp(x),x=0,5),polynom);    (Maple)
```

Exercises 6.1   *Hints, Suggestions, Solutions, and Examples*

3. By the ratio test,

$$\lim_{k\to\infty}\left|\frac{(x-5)^{k+1}/10^{k+1}}{(x-5)^k/10^k}\right| = \lim_{k\to\infty}\frac{1}{10}|x-5| = \frac{1}{10}|x-5|.$$

The series is absolutely convergent for $\frac{1}{10}|x - 5| < 1$, $|x - 5| < 10$, or on $(-5, 15)$. The radius of convergence is $R = 10$. At $x = -5$, the series $\sum_{k=1}^{\infty}(-1)^k(-10)^k/10^k = \sum_{k=1}^{\infty} 1$ diverges by the nth term test. At $x = 15$, the series $\sum_{k=1}^{\infty}(-1)^k 10^k/10^k = \sum_{k=1}^{\infty}(-1)^k$ diverges by the nth term test. Thus, the series converges on $(-5, 15)$.

6. $e^{-x}\cos x = \left(1 - x + \dfrac{x^2}{2} - \dfrac{x^3}{6} + \dfrac{x^4}{24} - \cdots\right)\left(1 - \dfrac{x^2}{2} + \dfrac{x^4}{24} - \cdots\right) = 1 - x + \dfrac{x^3}{3} - \dfrac{x^4}{6} + \cdots$

9. Let $k = n + 2$ so that $n = k - 2$ and

$$\sum_{n=1}^{\infty} nc_n x^{n+2} = \sum_{k=3}^{\infty}(k-2)c_{k-2}x^k.$$

12. $\displaystyle\sum_{n=2}^{\infty} n(n-1)c_n x^n + 2\sum_{n=2}^{\infty} n(n-1)c_n x^{n-2} + 3\sum_{n=1}^{\infty} nc_n x^n$

$$= 2 \cdot 2 \cdot 1c_2 x^0 + 2 \cdot 3 \cdot 2c_3 x^1 + 3 \cdot 1 \cdot c_1 x^1 + \underbrace{\sum_{n=2}^{\infty} n(n-1)c_n x^n}_{k=n} + 2\underbrace{\sum_{n=4}^{\infty} n(n-1)c_n x^{n-2}}_{k=n-2} + 3\underbrace{\sum_{n=2}^{\infty} nc_n x^n}_{k=n}$$

$$= 4c_2 + (3c_1 + 12c_3)x + \sum_{k=2}^{\infty} k(k-1)c_k x^k + 2\sum_{k=2}^{\infty}(k+2)(k+1)c_{k+2}x^k + 3\sum_{k=2}^{\infty} kc_k x^k$$

$$= 4c_2 + (3c_1 + 12c_3)x + \sum_{k=2}^{\infty}[(k(k-1) + 3k)c_k + 2(k+2)(k+1)c_{k+2}]x^k$$

$$= 4c_2 + (3c_1 + 12c_3)x + \sum_{k=2}^{\infty}[k(k+2)c_k + 2(k+1)(k+2)c_{k+2}]x^k$$

15. The singular points of $(x^2 - 25)y'' + 2xy' + y = 0$ are -5 and 5. The distance from 0 to either of these points is 5. The distance from 1 to the closest of these points is 4.

18. Substituting $y = \sum_{n=0}^{\infty} c_n x^n$ into the DE we have

$$y'' + x^2 y = \underbrace{\sum_{n=2}^{\infty} n(n-1)c_n x^{n-2}}_{k=n-2} + \underbrace{\sum_{n=0}^{\infty} c_n x^{n+2}}_{k=n+2} = \sum_{k=0}^{\infty}(k+2)(k+1)c_{k+2}x^k + \sum_{k=2}^{\infty} c_{k-2}x^k$$

$$= 2c_2 + 6c_3 x + \sum_{k=2}^{\infty}[(k+2)(k+1)c_{k+2} + c_{k-2}]x^k = 0.$$

Thus

$$c_2 = c_3 = 0$$

$$(k+2)(k+1)c_{k+2} + c_{k-2} = 0$$

and

$$c_{k+2} = -\frac{1}{(k+2)(k+1)}c_{k-2}, \quad k = 2, 3, 4, \ldots.$$

Choosing $c_0 = 1$ and $c_1 = 0$ we find

$$c_4 = -\frac{1}{12}$$

$$c_5 = c_6 = c_7 = 0$$

$$c_8 = \frac{1}{672}$$

and so on. For $c_0 = 0$ and $c_1 = 1$ we obtain

$$c_4 = 0$$

$$c_5 = -\frac{1}{20}$$

$$c_6 = c_7 = c_8 = 0$$

$$c_9 = \frac{1}{1440}$$

and so on. Thus, two solutions are

$$y_1 = 1 - \frac{1}{12}x^4 + \frac{1}{672}x^8 - \cdots \qquad \text{and} \qquad y_2 = x - \frac{1}{20}x^5 + \frac{1}{1440}x^9 - \cdots.$$

21. Substituting $y = \sum_{n=0}^{\infty} c_n x^n$ into the DE we have

$$y'' + x^2 y' + xy = \underbrace{\sum_{n=2}^{\infty} n(n-1)c_n x^{n-2}}_{k=n-2} + \underbrace{\sum_{n=1}^{\infty} n c_n x^{n+1}}_{k=n+1} + \underbrace{\sum_{n=0}^{\infty} c_n x^{n+1}}_{k=n+1}$$

$$= \sum_{k=0}^{\infty} (k+2)(k+1)c_{k+2} x^k + \sum_{k=2}^{\infty} (k-1)c_{k-1} x^k + \sum_{k=1}^{\infty} c_{k-1} x^k$$

$$= 2c_2 + (6c_3 + c_0)x + \sum_{k=2}^{\infty} [(k+2)(k+1)c_{k+2} + kc_{k-1}]x^k = 0.$$

Thus

$$c_2 = 0$$

$$6c_3 + c_0 = 0$$

$$(k+2)(k+1)c_{k+2} + kc_{k-1} = 0$$

and

$$c_2 = 0$$

$$c_3 = -\frac{1}{6}c_0$$

$$c_{k+2} = -\frac{k}{(k+2)(k+1)}c_{k-1}, \quad k = 2, 3, 4, \ldots.$$

Choosing $c_0 = 1$ and $c_1 = 0$ we find

$$c_3 = -\frac{1}{6}$$

$$c_4 = c_5 = 0$$

$$c_6 = \frac{1}{45}$$

and so on. For $c_0 = 0$ and $c_1 = 1$ we obtain

$$c_3 = 0$$

$$c_4 = -\frac{1}{6}$$

$$c_5 = c_6 = 0$$

$$c_7 = \frac{5}{252}$$

and so on. Thus, two solutions are

$$y_1 = 1 - \frac{1}{6}x^3 + \frac{1}{45}x^6 - \cdots \qquad \text{and} \qquad y_2 = x - \frac{1}{6}x^4 + \frac{5}{252}x^7 - \cdots .$$

24. Substituting $y = \sum_{n=0}^{\infty} c_n x^n$ into the DE we have

$$(x+2)y'' + xy' - y = \underbrace{\sum_{n=2}^{\infty} n(n-1)c_n x^{n-1}}_{k=n-1} + \underbrace{\sum_{n=2}^{\infty} 2n(n-1)c_n x^{n-2}}_{k=n-2} + \underbrace{\sum_{n=1}^{\infty} nc_n x^n}_{k=n} - \underbrace{\sum_{n=0}^{\infty} c_n x^n}_{k=n}$$

$$= \sum_{k=1}^{\infty} (k+1)kc_{k+1} x^k + \sum_{k=0}^{\infty} 2(k+2)(k+1)c_{k+2} x^k + \sum_{k=1}^{\infty} kc_k x^k - \sum_{k=0}^{\infty} c_k x^k$$

$$= 4c_2 - c_0 + \sum_{k=1}^{\infty} [(k+1)kc_{k+1} + 2(k+2)(k+1)c_{k+2} + (k-1)c_k]x^k = 0.$$

Thus

$$4c_2 - c_0 = 0$$

$$(k+1)kc_{k+1} + 2(k+2)(k+1)c_{k+2} + (k-1)c_k = 0, \quad k = 1, 2, 3, \ldots$$

and

$$c_2 = \frac{1}{4}c_0$$

$$c_{k+2} = -\frac{(k+1)kc_{k+1} + (k-1)c_k}{2(k+2)(k+1)}, \quad k = 1, 2, 3, \ldots .$$

Choosing $c_0 = 1$ and $c_1 = 0$ we find

$$c_1 = 0, \qquad c_2 = \frac{1}{4}, \qquad c_3 = -\frac{1}{24}, \qquad c_4 = 0, \qquad c_5 = \frac{1}{480}$$

and so on. For $c_0 = 0$ and $c_1 = 1$ we obtain

$$c_2 = 0$$

$$c_3 = 0$$

$$c_4 = c_5 = c_6 = \cdots = 0.$$

Thus, two solutions are

$$y_1 = c_0\left[1 + \frac{1}{4}x^2 - \frac{1}{24}x^3 + \frac{1}{480}x^5 + \cdots\right] \quad \text{and} \quad y_2 = c_1 x.$$

27. Substituting $y = \sum_{n=0}^{\infty} c_n x^n$ into the DE we have

$$\left(x^2 + 2\right)y'' + 3xy' - y = \underbrace{\sum_{n=2}^{\infty} n(n-1)c_n x^n}_{k=n} + 2\underbrace{\sum_{n=2}^{\infty} n(n-1)c_n x^{n-2}}_{k=n-2} + 3\underbrace{\sum_{n=1}^{\infty} nc_n x^n}_{k=n} - \underbrace{\sum_{n=0}^{\infty} c_n x^n}_{k=n}$$

$$= \sum_{k=2}^{\infty} k(k-1)c_k x^k + 2\sum_{k=0}^{\infty}(k+2)(k+1)c_{k+2} x^k + 3\sum_{k=1}^{\infty} kc_k x^k - \sum_{k=0}^{\infty} c_k x^k$$

$$= (4c_2 - c_0) + (12c_3 + 2c_1)x + \sum_{k=2}^{\infty}\left[2(k+2)(k+1)c_{k+2} + \left(k^2 + 2k - 1\right)c_k\right]x^k = 0.$$

Thus

$$4c_2 - c_0 = 0$$

$$12c_3 + 2c_1 = 0$$

$$2(k+2)(k+1)c_{k+2} + \left(k^2 + 2k - 1\right)c_k = 0$$

and

$$c_2 = \frac{1}{4}c_0$$

$$c_3 = -\frac{1}{6}c_1$$

$$c_{k+2} = -\frac{k^2 + 2k - 1}{2(k+2)(k+1)}c_k, \quad k = 2, 3, 4, \ldots.$$

Choosing $c_0 = 1$ and $c_1 = 0$ we find

$$c_2 = \frac{1}{4}$$

$$c_3 = c_5 = c_7 = \cdots = 0$$

$$c_4 = -\frac{7}{96}$$

and so on. For $c_0 = 0$ and $c_1 = 1$ we obtain

$$c_2 = c_4 = c_6 = \cdots = 0$$

$$c_3 = -\frac{1}{6}$$

$$c_5 = \frac{7}{120}$$

and so on. Thus, two solutions are

$$y_1 = 1 + \frac{1}{4}x^2 - \frac{7}{96}x^4 + \cdots \qquad \text{and} \qquad y_2 = x - \frac{1}{6}x^3 + \frac{7}{120}x^5 - \cdots .$$

30. Substituting $y = \sum_{n=0}^{\infty} c_n x^n$ into the DE we have

$$(x+1)y'' - (2-x)y' + y$$

$$= \underbrace{\sum_{n=2}^{\infty} n(n-1)c_n x^{n-1}}_{k=n-1} + \underbrace{\sum_{n=2}^{\infty} n(n-1)c_n x^{n-2}}_{k=n-2} - 2\underbrace{\sum_{n=1}^{\infty} nc_n x^{n-1}}_{k=n-1} + \underbrace{\sum_{n=1}^{\infty} nc_n x^{n}}_{k=n} + \underbrace{\sum_{n=0}^{\infty} c_n x^{n}}_{k=n}$$

$$= \sum_{k=1}^{\infty} (k+1)kc_{k+1}x^k + \sum_{k=0}^{\infty}(k+2)(k+1)c_{k+2}x^k - 2\sum_{k=0}^{\infty}(k+1)c_{k+1}x^k + \sum_{k=1}^{\infty} kc_k x^k + \sum_{k=0}^{\infty} c_k x^k$$

$$= 2c_2 - 2c_1 + c_0 + \sum_{k=1}^{\infty}[(k+2)(k+1)c_{k+2} - (k+1)c_{k+1} + (k+1)c_k]x^k = 0.$$

Thus

$$2c_2 - 2c_1 + c_0 = 0$$

$$(k+2)(k+1)c_{k+2} - (k+1)c_{k+1} + (k+1)c_k = 0$$

and

$$c_2 = c_1 - \frac{1}{2}c_0$$

$$c_{k+2} = \frac{1}{k+2}c_{k+1} - \frac{1}{k+2}c_k, \quad k = 1, 2, 3, \ldots .$$

Choosing $c_0 = 1$ and $c_1 = 0$ we find

$$c_2 = -\frac{1}{2}, \qquad c_3 = -\frac{1}{6}, \qquad c_4 = \frac{1}{12},$$

and so on. For $c_0 = 0$ and $c_1 = 1$ we obtain

$$c_2 = 1, \qquad c_3 = 0, \qquad c_4 = -\frac{1}{4},$$

and so on. Thus,

$$y = C_1\left(1 - \frac{1}{2}x^2 - \frac{1}{6}x^3 + \frac{1}{12}x^4 + \cdots\right) + C_2\left(x + x^2 - \frac{1}{4}x^4 + \cdots\right)$$

and

$$y' = C_1 \left(-x - \frac{1}{2}x^2 + \frac{1}{3}x^3 + \cdots \right) + C_2 \left(1 + 2x - x^3 + \cdots \right).$$

The initial conditions imply $C_1 = 2$ and $C_2 = -1$, so

$$y = 2 \left(1 - \frac{1}{2}x^2 - \frac{1}{6}x^3 + \frac{1}{12}x^4 + \cdots \right) - \left(x + x^2 - \frac{1}{4}x^4 + \cdots \right)$$

$$= 2 - x - 2x^2 - \frac{1}{3}x^3 + \frac{5}{12}x^4 + \cdots.$$

33. Substituting $y = \sum_{n=0}^{\infty} c_n x^n$ into the DE we have

$$y'' + (\sin x)y = \sum_{n=2}^{\infty} n(n-1)c_n x^{n-2} + \left(x - \frac{1}{6}x^3 + \frac{1}{120}x^5 - \cdots \right) \left(c_0 + c_1 x + c_2 x^2 + \cdots \right)$$

$$= \left[2c_2 + 6c_3 x + 12c_4 x^2 + 20c_5 x^3 + \cdots \right] + \left[c_0 x + c_1 x^2 + \left(c_2 - \frac{1}{6}c_0 \right) x^3 + \cdots \right]$$

$$= 2c_2 + (6c_3 + c_0)x + (12c_4 + c_1)x^2 + \left(20c_5 + c_2 - \frac{1}{6}c_0 \right) x^3 + \cdots = 0.$$

Thus

$$2c_2 = 0$$

$$6c_3 + c_0 = 0$$

$$12c_4 + c_1 = 0$$

$$20c_5 + c_2 - \frac{1}{6}c_0 = 0$$

and

$$c_2 = 0$$

$$c_3 = -\frac{1}{6}c_0$$

$$c_4 = -\frac{1}{12}c_1$$

$$c_5 = -\frac{1}{20}c_2 + \frac{1}{120}c_0.$$

Choosing $c_0 = 1$ and $c_1 = 0$ we find

$$c_2 = 0, \qquad c_3 = -\frac{1}{6}, \qquad c_4 = 0, \qquad c_5 = \frac{1}{120}$$

and so on. For $c_0 = 0$ and $c_1 = 1$ we obtain

$$c_2 = 0, \qquad c_3 = 0, \qquad c_4 = -\frac{1}{12}, \qquad c_5 = 0$$

and so on. Thus, two solutions are

$$y_1 = 1 - \frac{1}{6}x^3 + \frac{1}{120}x^5 + \cdots \qquad \text{and} \qquad y_2 = x - \frac{1}{12}x^4 + \cdots.$$

Section 6.2   Solutions About Singular Points

The terminology and concepts listed below provide an outline of the main ideas encountered in this section. These can be useful when preparing for a quiz or test.

Terminology and Concepts

- regular singular point of a DE
- irregular singular point of a DE
- method of Frobenius for finding a series solution about a regular singular point of a DE
- indicial equation

The basic skills listed below summarize the more mechanical types of problems encountered in the exercise set for this section.

Basic Skills

- determine the singular points of a linear DE and classify them as regular or irregular
- use the method of Frobenius to obtain two linearly independent solutions about $x = 0$ of a homogeneous linear DE with regular singular point $x = 0$ when the indicial roots
  1. do not differ by an integer, or
  2. differ by an integer.

Exercises 6.2   *Hints, Suggestions, Solutions, and Examples*

3. Irregular singular point: $x = 3$; regular singular point: $x = -3$

6. Irregular singular point: $x = 5$; regular singular point: $x = 0$

9. Irregular singular point: $x = 0$; regular singular points: $x = 2, \pm 5$

12. Writing the DE in the form

$$y'' + \frac{x+3}{x}\, y' + 7xy = 0$$

we see that $x_0 = 0$ is a regular singular point. Multiplying by x^2, the DE can be put in the form

$$x^2 y'' + x(x+3)y' + 7x^3 y = 0.$$

We identify $p(x) = x + 3$ and $q(x) = 7x^3$.

15. Substituting $y = \sum_{n=0}^{\infty} c_n x^{n+r}$ into the DE and collecting terms, we obtain

$$2xy'' - y' + 2y = \left(2r^2 - 3r\right) c_0 x^{r-1} + \sum_{k=1}^{\infty} [2(k+r-1)(k+r)c_k - (k+r)c_k + 2c_{k-1}]x^{k+r-1} = 0,$$

which implies

$$2r^2 - 3r = r(2r - 3) = 0$$

and

$$(k+r)(2k + 2r - 3)c_k + 2c_{k-1} = 0.$$

The indicial roots are $r = 0$ and $r = 3/2$. For $r = 0$ the recurrence relation is

$$c_k = -\frac{2c_{k-1}}{k(2k-3)}, \quad k = 1, 2, 3, \ldots,$$

and

$$c_1 = 2c_0, \qquad c_2 = -2c_0, \qquad c_3 = \frac{4}{9}c_0,$$

and so on. For $r = 3/2$ the recurrence relation is

$$c_k = -\frac{2c_{k-1}}{(2k+3)k}, \quad k = 1, 2, 3, \ldots,$$

and

$$c_1 = -\frac{2}{5}c_0, \qquad c_2 = \frac{2}{35}c_0, \qquad c_3 = -\frac{4}{945}c_0,$$

and so on. The general solution on $(0, \infty)$ is

$$y = C_1 \left(1 + 2x - 2x^2 + \frac{4}{9}x^3 + \cdots\right) + C_2 x^{3/2}\left(1 - \frac{2}{5}x + \frac{2}{35}x^2 - \frac{4}{945}x^3 + \cdots\right).$$

18. Substituting $y = \sum_{n=0}^{\infty} c_n x^{n+r}$ into the DE and collecting terms, we obtain

$$2x^2 y'' - xy' + \left(x^2 + 1\right)y = \left(2r^2 - 3r + 1\right)c_0 x^r + \left(2r^2 + r\right)c_1 x^{r+1}$$

$$+ \sum_{k=2}^{\infty} [2(k+r)(k+r-1)c_k - (k+r)c_k + c_k + c_{k-2}]x^{k+r}$$

$$= 0,$$

which implies

$$2r^2 - 3r + 1 = (2r - 1)(r - 1) = 0,$$

$$\left(2r^2 + r\right)c_1 = 0,$$

131

and
$$[(k+r)(2k+2r-3)+1]c_k + c_{k-2} = 0.$$

The indicial roots are $r = 1/2$ and $r = 1$, so $c_1 = 0$. For $r = 1/2$ the recurrence relation is

$$c_k = -\frac{c_{k-2}}{k(2k-1)}, \quad k = 2,3,4,\ldots,$$

and
$$c_2 = -\frac{1}{6}c_0, \quad c_3 = 0, \quad c_4 = \frac{1}{168}c_0,$$

and so on. For $r = 1$ the recurrence relation is

$$c_k = -\frac{c_{k-2}}{k(2k+1)}, \quad k = 2,3,4,\ldots,$$

and
$$c_2 = -\frac{1}{10}c_0, \quad c_3 = 0, \quad c_4 = \frac{1}{360}c_0,$$

and so on. The general solution on $(0,\infty)$ is

$$y = C_1 x^{1/2}\left(1 - \frac{1}{6}x^2 + \frac{1}{168}x^4 + \cdots\right) + C_2 x\left(1 - \frac{1}{10}x^2 + \frac{1}{360}x^4 + \cdots\right).$$

21. Substituting $y = \sum_{n=0}^{\infty} c_n x^{n+r}$ into the DE and collecting terms, we obtain

$$2xy'' - (3+2x)y' + y = \left(2r^2 - 5r\right)c_0 x^{r-1} + \sum_{k=1}^{\infty}[2(k+r)(k+r-1)c_k$$

$$- 3(k+r)c_k - 2(k+r-1)c_{k-1} + c_{k-1}]x^{k+r-1}$$

$$= 0,$$

which implies

$$2r^2 - 5r = r(2r-5) = 0$$

and
$$(k+r)(2k+2r-5)c_k - (2k+2r-3)c_{k-1} = 0.$$

The indicial roots are $r = 0$ and $r = 5/2$. For $r = 0$ the recurrence relation is

$$c_k = \frac{(2k-3)c_{k-1}}{k(2k-5)}, \quad k = 1,2,3,\ldots,$$

and
$$c_1 = \frac{1}{3}c_0, \quad c_2 = -\frac{1}{6}c_0, \quad c_3 = -\frac{1}{6}c_0,$$

and so on. For $r = 5/2$ the recurrence relation is

$$c_k = \frac{2(k+1)c_{k-1}}{k(2k+5)}, \quad k = 1,2,3,\ldots,$$

and

$$c_1 = \frac{4}{7}c_0, \qquad c_2 = \frac{4}{21}c_0, \qquad c_3 = \frac{32}{693}c_0,$$

and so on. The general solution on $(0, \infty)$ is

$$y = C_1\left(1 + \frac{1}{3}x - \frac{1}{6}x^2 - \frac{1}{6}x^3 + \cdots\right) + C_2 x^{5/2}\left(1 + \frac{4}{7}x + \frac{4}{21}x^2 + \frac{32}{693}x^3 + \cdots\right).$$

24. Substituting $y = \sum_{n=0}^{\infty} c_n x^{n+r}$ into the DE and collecting terms, we obtain

$$2x^2 y'' + 3xy' + (2x - 1)y = \left(2r^2 + r - 1\right)c_0 x^r$$

$$+ \sum_{k=1}^{\infty}[2(k+r)(k+r-1)c_k + 3(k+r)c_k - c_k + 2c_{k-1}]x^{k+r}$$

$$= 0,$$

which implies

$$2r^2 + r - 1 = (2r-1)(r+1) = 0$$

and

$$[(k+r)(2k+2r+1) - 1]c_k + 2c_{k-1} = 0.$$

The indicial roots are $r = -1$ and $r = 1/2$. For $r = -1$ the recurrence relation is

$$c_k = -\frac{2c_{k-1}}{k(2k-3)}, \quad k = 1, 2, 3, \ldots,$$

and

$$c_1 = 2c_0, \qquad c_2 = -2c_0, \qquad c_3 = \frac{4}{9}c_0,$$

and so on. For $r = 1/2$ the recurrence relation is

$$c_k = -\frac{2c_{k-1}}{k(2k+3)}, \quad k = 1, 2, 3, \ldots,$$

and

$$c_1 = -\frac{2}{5}c_0, \qquad c_2 = \frac{2}{35}c_0, \qquad c_3 = -\frac{4}{945}c_0,$$

and so on. The general solution on $(0, \infty)$ is

$$y = C_1 x^{-1}\left(1 + 2x - 2x^2 + \frac{4}{9}x^3 + \cdots\right) + C_2 x^{1/2}\left(1 - \frac{2}{5}x + \frac{2}{35}x^2 - \frac{4}{945}x^3 + \cdots\right).$$

27. Substituting $y = \sum_{n=0}^{\infty} c_n x^{n+r}$ into the DE and collecting terms, we obtain

$$xy'' - xy' + y = \left(r^2 - r\right)c_0 x^{r-1} + \sum_{k=0}^{\infty}[(k+r+1)(k+r)c_{k+1} - (k+r)c_k + c_k]x^{k+r} = 0$$

which implies

$$r^2 - r = r(r-1) = 0$$

and

$$(k+r+1)(k+r)c_{k+1} - (k+r-1)c_k = 0.$$

The indicial roots are $r_1 = 1$ and $r_2 = 0$. For $r_1 = 1$ the recurrence relation is

$$c_{k+1} = \frac{kc_k}{(k+2)(k+1)}, \quad k = 0, 1, 2, \ldots,$$

and one solution is $y_1 = c_0 x$. A second solution is

$$y_2 = x\int \frac{e^{-\int -1\,dx}}{x^2}\,dx = x\int \frac{e^x}{x^2}\,dx = x\int \frac{1}{x^2}\left(1 + x + \frac{1}{2}x^2 + \frac{1}{3!}x^3 + \cdots\right)dx$$

$$= x\int\left(\frac{1}{x^2} + \frac{1}{x} + \frac{1}{2} + \frac{1}{3!}x + \frac{1}{4!}x^2 + \cdots\right)dx = x\left[-\frac{1}{x} + \ln x + \frac{1}{2}x + \frac{1}{12}x^2 + \frac{1}{72}x^3 + \cdots\right]$$

$$= x\ln x - 1 + \frac{1}{2}x^2 + \frac{1}{12}x^3 + \frac{1}{72}x^4 + \cdots.$$

The general solution on $(0, \infty)$ is

$$y = C_1 x + C_2 y_2(x).$$

28. (*Example*) Suppose a power series $\sum_{n=0}^{\infty} c_k x^k$ is determined by the recurrence relation

$$c_k = c_{k-2}/(k+1), \quad c_0 = 1, \ c_1 = 0.$$

Find the power series representation of $1/(\sum_{k=0}^{\infty} c_k x^k)^2$.

```
Clear[c]                                              (Mathematica)
c[k_]:=c[k-2]/(k+1)
c[0]=1; c[1]=0;
ser=Sum[c[k]x^k, {k, 0, 6}]+0[x]^7
1/ser^2
```

```
with(powseries);                                          (Maple)
c:='c':
powcreate(c(k)=c(k-2)/(k+1),c(0)=1,c(1)=0):
inv=inverse(multiply(c,c)):
tpsform(inv,x,7);
```

30. Substituting $y = \sum_{n=0}^{\infty} c_n x^{n+r}$ into the DE and collecting terms, we obtain

$$xy'' + y' + y = r^2 c_0 x^{r-1} + \sum_{k=1}^{\infty}[(k+r)(k+r-1)c_k + (k+r)c_k + c_{k-1}]x^{k+r-1} = 0$$

which implies $r^2 = 0$ and

$$(k+r)^2 c_k + c_{k-1} = 0.$$

The indicial roots are $r_1 = r_2 = 0$ and the recurrence relation is

$$c_k = -\frac{c_{k-1}}{k^2}, \quad k = 1, 2, 3, \ldots .$$

One solution is

$$y_1 = c_0 \left(1 - x + \frac{1}{2^2}x^2 - \frac{1}{(3!)^2}x^3 + \frac{1}{(4!)^2}x^4 - \cdots\right) = c_0 \sum_{n=0}^{\infty} \frac{(-1)^n}{(n!)^2}x^n.$$

A second solution is

$$y_2 = y_1 \int \frac{e^{-\int (1/x)dx}}{y_1^2} \, dx = y_1 \int \frac{dx}{x\left(1 - x + \frac{1}{4}x^2 - \frac{1}{36}x^3 + \cdots\right)^2}$$

$$= y_1 \int \frac{dx}{x\left(1 - 2x + \frac{3}{2}x^2 - \frac{5}{9}x^3 + \frac{35}{288}x^4 - \cdots\right)}$$

$$= y_1 \int \frac{1}{x}\left(1 + 2x + \frac{5}{2}x^2 + \frac{23}{9}x^3 + \frac{677}{288}x^4 + \cdots\right) dx$$

$$= y_1 \int \left(\frac{1}{x} + 2 + \frac{5}{2}x + \frac{23}{9}x^2 + \frac{677}{.288}x^3 + \cdots\right) dx$$

$$= y_1 \left[\ln x + 2x + \frac{5}{4}x^2 + \frac{23}{27}x^3 + \frac{677}{1,152}x^4 + \cdots\right]$$

$$= y_1 \ln x + y_1 \left(2x + \frac{5}{4}x^2 + \frac{23}{27}x^3 + \frac{677}{1,152}x^4 + \cdots\right).$$

The general solution on $(0, \infty)$ is

$$y = C_1 y_1(x) + C_2 y_2(x).$$

33. (a) From $t = 1/x$ we have $dt/dx = -1/x^2 = -t^2$. Then

$$\frac{dy}{dx} = \frac{dy}{dt}\frac{dt}{dx} = -t^2\frac{dy}{dt}$$

and

$$\frac{d^2y}{dx^2} = \frac{d}{dx}\left(\frac{dy}{dx}\right) = \frac{d}{dx}\left(-t^2\frac{dy}{dt}\right) = -t^2\frac{d^2y}{dt^2}\frac{dt}{dx} - \frac{dy}{dt}\left(2t\frac{dt}{dx}\right) = t^4\frac{d^2y}{dt^2} + 2t^3\frac{dy}{dt}.$$

Now

$$x^4\frac{d^2y}{dx^2} + \lambda y = \frac{1}{t^4}\left(t^4\frac{d^2y}{dt^2} + 2t^3\frac{dy}{dt}\right) + \lambda y = \frac{d^2y}{dt^2} + \frac{2}{t}\frac{dy}{dt} + \lambda y = 0$$

becomes

$$t\frac{d^2y}{dt^2} + 2\frac{dy}{dt} + \lambda t y = 0.$$

(b) Substituting $y = \sum_{n=0}^{\infty} c_n t^{n+r}$ into the DE and collecting terms, we obtain

$$t \frac{d^2 y}{dt^2} + 2 \frac{dy}{dt} + \lambda t y = (r^2 + r)c_0 t^{r-1} + (r^2 + 3r + 2)c_1 t^r$$

$$+ \sum_{k=2}^{\infty} [(k+r)(k+r-1)c_k + 2(k+r)c_k + \lambda c_{k-2}]t^{k+r-1}$$

$$= 0,$$

which implies

$$r^2 + r = r(r+1) = 0,$$

$$\left(r^2 + 3r + 2\right) c_1 = 0,$$

and

$$(k+r)(k+r+1)c_k + \lambda c_{k-2} = 0.$$

The indicial roots are $r_1 = 0$ and $r_2 = -1$, so $c_1 = 0$. For $r_1 = 0$ the recurrence relation is

$$c_k = -\frac{\lambda c_{k-2}}{k(k+1)}, \quad k = 2, 3, 4, \ldots,$$

and

$$c_2 = -\frac{\lambda}{3!} c_0$$

$$c_3 = c_5 = c_7 = \cdots = 0$$

$$c_4 = \frac{\lambda^2}{5!} c_0$$

$$\vdots$$

$$c_{2n} = (-1)^n \frac{\lambda^n}{(2n+1)!} c_0.$$

For $r_2 = -1$ the recurrence relation is

$$c_k = -\frac{\lambda c_{k-2}}{k(k-1)}, \quad k = 2, 3, 4, \ldots,$$

and

$$c_2 = -\frac{\lambda}{2!}c_0$$

$$c_3 = c_5 = c_7 = \cdots = 0$$

$$c_4 = \frac{\lambda^2}{4!}c_0$$

$$\vdots$$

$$c_{2n} = (-1)^n \frac{\lambda^n}{(2n)!}c_0.$$

The general solution on $(0, \infty)$ is

$$y(t) = c_1 \sum_{n=0}^{\infty} \frac{(-1)^n}{(2n+1)!}(\sqrt{\lambda}\,t)^{2n} + c_2 t^{-1} \sum_{n=0}^{\infty} \frac{(-1)^n}{(2n)!}(\sqrt{\lambda}\,t)^{2n}$$

$$= \frac{1}{t}\left[C_1 \sum_{n=0}^{\infty} \frac{(-1)^n}{(2n+1)!}(\sqrt{\lambda}\,t)^{2n+1} + C_2 \sum_{n=0}^{\infty} \frac{(-1)^n}{(2n)!}(\sqrt{\lambda}\,t)^{2n}\right]$$

$$= \frac{1}{t}[C_1 \sin\sqrt{\lambda}\,t + C_2 \cos\sqrt{\lambda}\,t].$$

(c) Using $t = 1/x$, the solution of the original equation is

$$y(x) = C_1 x \sin\frac{\sqrt{\lambda}}{x} + C_2 x \cos\frac{\sqrt{\lambda}}{x}.$$

Section 6.3 Special Functions

The terminology and concepts listed below provide an outline of the main ideas encountered in this section. These can be useful when preparing for a quiz or test.

Terminology and Concepts

- Bessel's DE of order ν
- Bessel functions of the first kind: $J_\nu(x)$ and $J_{-\nu}(x)$
- Bessel functions of the second kind: $Y_\nu(x)$
- parametric Bessel equation of order ν
- modified Bessel equation of order ν

- modified Bessel function of the first kind: $I_\nu(x)$
- modified Bessel function of the second kind: $K_\nu(x)$
- aging spring
- Euler's constant: γ
- differential recurrence relation
- spherical Bessel functions
- Legendre's DE of order n
- Legendre polynomials
- Rodrigues' formula

The basic skills listed below summarize the more mechanical types of problems encountered in the exercise set for this section.

Basic Skills

- solve Bessel's DE of order ν
- solve DEs of the form $x^2 y'' + xy' + (\alpha^2 x^2 - \nu^2)y = 0$
- solve DEs of the form $y'' + [(1 - 2a)/x]y' + [b^2 c^2 x^{2c-2} + (a^2 - p^2 c^2)/x^2]y = 0, \; p \geq 0$
- verify various recurrence relations involving Bessel functions
- use a recurrence relation to generate Legendre polynomials
- use Rodrigues' formula to generate Legendre polynomials

Use of Computers   See Section 2.3 in this manual for the syntax used to obtain Bessel functions and Legendre polynomials in *Mathematica* and *Maple*.

Exercises 6.3   *Hints, Suggestions, Solutions, and Examples*

3. Since $\nu^2 = 25/4$ the general solution is $y = c_1 J_{5/2}(x) + c_2 J_{-5/2}(x)$.

6. Since $\nu^2 = 4$ the general solution is $y = c_1 J_2(x) + c_2 Y_2(x)$.

9. We identify $\alpha = 5$ and $\nu = \frac{2}{3}$. Then the general solution is $y = c_1 J_{2/3}(5x) + c_2 J_{-2/3}(5x)$.

12. If $y = \sqrt{x}\, v(x)$ then

$$y' = x^{1/2} v'(x) + \frac{1}{2} x^{-1/2} v(x)$$

$$y'' = x^{1/2} v''(x) + x^{-1/2} v'(x) - \frac{1}{4} x^{-3/2} v(x)$$

and

$$x^2 y'' + \left(\alpha^2 x^2 - \nu^2 + \frac{1}{4}\right) y = x^{5/2} v''(x) + x^{3/2} v'(x) - \frac{1}{4} x^{1/2} v(x) + \left(\alpha^2 x^2 - \nu^2 + \frac{1}{4}\right) x^{1/2} v(x)$$

$$= x^{5/2} v''(x) + x^{3/2} v'(x) + (\alpha^2 x^{5/2} - \nu^2 x^{1/2}) v(x) = 0.$$

Multiplying by $x^{-1/2}$ we obtain

$$x^2 v''(x) + x v'(x) + (\alpha^2 x^2 - \nu^2) v(x) = 0,$$

whose solution is $v(x) = c_1 J_\nu(\alpha x) + c_2 Y_\nu(\alpha x)$. Then $y = c_1 \sqrt{x}\, J_\nu(\alpha x) + c_2 \sqrt{x}\, Y_\nu(\alpha x)$.

15. Write the DE in the form $y'' - (1/x)y' + y = 0$. This is the form of (18) in the text with $a = 1$, $c = 1$, $b = 1$, and $p = 1$, so, by (19) in the text, the general solution is

$$y = x[c_1 J_1(x) + c_2 Y_1(x)].$$

18. Write the DE in the form $y'' + (4 + 1/4x^2)y = 0$. This is the form of (18) in the text with $a = \frac{1}{2}$, $c = 1$, $b = 2$, and $p = 0$, so, by (19) in the text, the general solution is

$$y = x^{1/2}[c_1 J_0(2x) + c_2 Y_0(2x)].$$

21. Using the fact that $i^2 = -1$, along with the definition of $J_\nu(x)$ in (7) in the text, we have

$$I_\nu(x) = i^{-\nu} J_\nu(ix) = i^{-\nu} \sum_{n=0}^{\infty} \frac{(-1)^n}{n!\Gamma(1+\nu+n)} \left(\frac{ix}{2}\right)^{2n+\nu}$$

$$= \sum_{n=0}^{\infty} \frac{(-1)^n}{n!\Gamma(1+\nu+n)} i^{2n+\nu-\nu} \left(\frac{x}{2}\right)^{2n+\nu}$$

$$= \sum_{n=0}^{\infty} \frac{(-1)^n}{n!\Gamma(1+\nu+n)} (i^2)^n \left(\frac{x}{2}\right)^{2n+\nu}$$

$$= \sum_{n=0}^{\infty} \frac{(-1)^{2n}}{n!\Gamma(1+\nu+n)} \left(\frac{x}{2}\right)^{2n+\nu}$$

$$= \sum_{n=0}^{\infty} \frac{1}{n!\Gamma(1+\nu+n)} \left(\frac{x}{2}\right)^{2n+\nu},$$

which is a real function.

24. Write the DE in the form $y'' + (4/x)y' + (1 + 2/x^2)y = 0$. This is the form of (18) in the text with

$$1 - 2a = 4 \implies a = -\frac{3}{2}$$

$$2c - 2 = 0 \implies c = 1$$

$$b^2 c^2 = 1 \implies b = 1$$

$$a^2 - p^2 c^2 = 2 \implies p = \frac{1}{2}.$$

Then, by (19), (23), and (24) in the text,

$$y = x^{-3/2}[c_1 J_{1/2}(x) + c_2 J_{-1/2}(x)] = x^{-3/2}\left[c_1\sqrt{\frac{2}{\pi x}}\sin x + c_2\sqrt{\frac{2}{\pi x}}\cos x\right]$$

$$= C_1\frac{1}{x^2}\sin x + C_2\frac{1}{x^2}\cos x.$$

27. (a) The recurrence relation follows from

$$-\nu J_\nu(x) + x J_{\nu-1}(x) = -\sum_{n=0}^{\infty}\frac{(-1)^n \nu}{n!\Gamma(1+\nu+n)}\left(\frac{x}{2}\right)^{2n+\nu} + x\sum_{n=0}^{\infty}\frac{(-1)^n}{n!\Gamma(\nu+n)}\left(\frac{x}{2}\right)^{2n+\nu-1}$$

$$= -\sum_{n=0}^{\infty}\frac{(-1)^n \nu}{n!\Gamma(1+\nu+n)}\left(\frac{x}{2}\right)^{2n+\nu} + \sum_{n=0}^{\infty}\frac{(-1)^n(\nu+n)}{n!\Gamma(1+\nu+n)}\cdot 2\left(\frac{x}{2}\right)\left(\frac{x}{2}\right)^{2n+\nu-1}$$

$$= \sum_{n=0}^{\infty}\frac{(-1)^n(2n+\nu)}{n!\Gamma(1+\nu+n)}\left(\frac{x}{2}\right)^{2n+\nu} = x J_\nu'(x).$$

(b) The formula in part (a) is a linear first-order DE in $J_\nu(x)$. An integrating factor for this equation is x^ν, so

$$\frac{d}{dx}[x^\nu J_\nu(x)] = x^\nu J_{\nu-1}(x).$$

30. From (20), $J_0'(x) = -J_1(x)$, and from (21), $J_0'(x) = J_{-1}(x)$. Thus $J_0'(x) = J_{-1}(x) = -J_1(x)$.

33. Letting

$$s = \frac{2}{\alpha}\sqrt{\frac{k}{m}}\,e^{-\alpha t/2},$$

we have

$$\frac{dx}{dt} = \frac{dx}{ds}\frac{ds}{dt} = \frac{dx}{dt}\left[\frac{2}{\alpha}\sqrt{\frac{k}{m}}\left(-\frac{\alpha}{2}\right)e^{-\alpha t/2}\right] = \frac{dx}{ds}\left(-\sqrt{\frac{k}{m}}\,e^{-\alpha t/2}\right)$$

and

$$\frac{d^2x}{dt^2} = \frac{d}{dt}\left(\frac{dx}{dt}\right) = \frac{dx}{ds}\left(\frac{\alpha}{2}\sqrt{\frac{k}{m}}\,e^{-\alpha t/2}\right) + \frac{d}{dt}\left(\frac{dx}{ds}\right)\left(-\sqrt{\frac{k}{m}}\,e^{-\alpha t/2}\right)$$

$$= \frac{dx}{ds}\left(\frac{\alpha}{2}\sqrt{\frac{k}{m}}\,e^{-\alpha t/2}\right) + \frac{d^2x}{ds^2}\frac{ds}{dt}\left(-\sqrt{\frac{k}{m}}\,e^{-\alpha t/2}\right)$$

$$= \frac{dx}{ds}\left(\frac{\alpha}{2}\sqrt{\frac{k}{m}}\,e^{-\alpha t/2}\right) + \frac{d^2x}{ds^2}\left(\frac{k}{m}\,e^{-\alpha t}\right).$$

Then

$$m\frac{d^2x}{dt^2} + ke^{-\alpha t}x = ke^{-\alpha t}\frac{d^2x}{ds^2} + \frac{m\alpha}{2}\sqrt{\frac{k}{m}}\,e^{-\alpha t/2}\frac{dx}{ds} + ke^{-\alpha t}x = 0.$$

Multiplying by $2^2/\alpha^2 m$ we have

$$\frac{2^2}{\alpha^2}\frac{k}{m}e^{-\alpha t}\frac{d^2x}{ds^2} + \frac{2}{\alpha}\sqrt{\frac{k}{m}}\,e^{-\alpha t/2}\frac{dx}{ds} + \frac{2^2}{\alpha^2}\frac{k}{m}e^{-\alpha t}x = 0$$

or, since $s = (2/\alpha)\sqrt{k/m}\,e^{-\alpha t/2}$,

$$s^2\frac{d^2x}{ds^2} + s\frac{dx}{ds} + s^2x = 0.$$

36. The general solution of the DE is

$$y(x) = c_1 J_0(\alpha x) + c_2 Y_0(\alpha x).$$

In order to satisfy the conditions that $\lim_{x\to 0+} y(x)$ and $\lim_{x\to 0+} y'(x)$ are finite we are forced to define $c_2 = 0$. Thus, $y(x) = c_1 J_0(\alpha x)$. The second boundary condition, $y(2) = 0$, implies $c_1 = 0$ or $J_0(2\alpha) = 0$. In order to have a nontrivial solution we require that $J_0(2\alpha) = 0$. From Table 6.1, the first three positive zeros of J_0 are found to be

$$2\alpha_1 = 2.4048, \quad 2\alpha_2 = 5.5201, \quad 2\alpha_3 = 8.6537$$

and so $\alpha_1 = 1.2024$, $\alpha_2 = 2.7601$, $\alpha_3 = 4.3269$. The eigenfunctions corresponding to the eigenvalues $\lambda_1 = \alpha_1^2$, $\lambda_2 = \alpha_2^2$, $\lambda_3 = \alpha_3^2$ are $J_0(1.2024x)$, $J_0(2.7601x)$, and $J_0(4.3269x)$.

38. (*Suggestion*) See the table of special functions in Section 2.3 of this manual for syntax in *Mathematica* and *Maple* for the modified Bessel functions.

45. The recurrence relation can be written

$$P_{k+1}(x) = \frac{2k+1}{k+1}xP_k(x) - \frac{k}{k+1}P_{k-1}(x), \qquad k = 2,\ 3,\ 4,\ \ldots\ .$$

$k = 1: \quad P_2(x) = \dfrac{3}{2}x^2 - \dfrac{1}{2}$

$k = 2: \quad P_3(x) = \dfrac{5}{3}x\left(\dfrac{3}{2}x^2 - \dfrac{1}{2}\right) - \dfrac{2}{3}x = \dfrac{5}{2}x^3 - \dfrac{3}{2}x$

$k = 3: \quad P_4(x) = \dfrac{7}{4}x\left(\dfrac{5}{2}x^3 - \dfrac{3}{2}x\right) - \dfrac{3}{4}\left(\dfrac{3}{2}x^2 - \dfrac{1}{2}\right) = \dfrac{35}{8}x^4 - \dfrac{30}{8}x^2 + \dfrac{3}{8}$

$k = 4: \quad P_5(x) = \dfrac{9}{5}x\left(\dfrac{35}{8}x^4 - \dfrac{30}{8}x^2 + \dfrac{3}{8}\right) - \dfrac{4}{5}\left(\dfrac{5}{2}x^3 - \dfrac{3}{2}x\right) = \dfrac{63}{8}x^5 - \dfrac{35}{4}x^3 + \dfrac{15}{8}x$

$k = 5: \quad P_6(x) = \dfrac{11}{6}x\left(\dfrac{63}{8}x^5 - \dfrac{35}{4}x^3 + \dfrac{15}{8}x\right) - \dfrac{5}{6}\left(\dfrac{35}{8}x^4 - \dfrac{30}{8}x^2 + \dfrac{3}{8}\right) = \dfrac{231}{16}x^6 - \dfrac{315}{16}x^4 + \dfrac{105}{16}x^2 - \dfrac{5}{16}$

$k = 6: \quad P_7(x) = \dfrac{13}{7}x\left(\dfrac{231}{16}x^6 - \dfrac{315}{16}x^4 + \dfrac{105}{16}x^2 - \dfrac{5}{16}\right) - \dfrac{6}{7}\left(\dfrac{63}{8}x^5 - \dfrac{35}{4}x^3 + \dfrac{15}{8}x\right)$

$$= \dfrac{429}{16}x^7 - \dfrac{693}{16}x^5 + \dfrac{315}{16}x^3 - \dfrac{35}{16}x$$

3. $x = -1$ is the nearest singular point to the ordinary point $x = 0$. Theorem 6.1.1 guarantees the existence of two power series solutions $y = \sum_{n=1}^{\infty} c_n x^n$ of the DE that converge at least for $-1 < x < 1$. Since $-\frac{1}{2} \le x \le \frac{1}{2}$ is properly contained in $-1 < x < 1$, both power series must converge for all points contained in $-\frac{1}{2} \le x \le \frac{1}{2}$.

6. We have

$$f(x) = \frac{\sin x}{\cos x} = \frac{x - \dfrac{x^3}{6} + \dfrac{x^5}{120} - \cdots}{1 - \dfrac{x^2}{2} + \dfrac{x^4}{24} - \cdots} = x + \frac{x^3}{3} + \frac{2x^5}{15} + \cdots.$$

9. Substituting $y = \sum_{n=0}^{\infty} c_n x^{n+r}$ into the DE we obtain

$$2xy'' + y' + y = \left(2r^2 - r\right) c_0 x^{r-1} + \sum_{k=1}^{\infty} [2(k+r)(k+r-1)c_k + (k+r)c_k + c_{k-1}]x^{k+r-1} = 0$$

which implies

$$2r^2 - r = r(2r - 1) = 0$$

and

$$(k+r)(2k+2r-1)c_k + c_{k-1} = 0.$$

The indicial roots are $r = 0$ and $r = 1/2$. For $r = 0$ the recurrence relation is

$$c_k = -\frac{c_{k-1}}{k(2k-1)}, \quad k = 1, 2, 3, \ldots,$$

so

$$c_1 = -c_0, \qquad c_2 = \frac{1}{6}c_0, \qquad c_3 = -\frac{1}{90}c_0.$$

For $r = 1/2$ the recurrence relation is

$$c_k = -\frac{c_{k-1}}{k(2k+1)}, \quad k = 1, 2, 3, \ldots,$$

so

$$c_1 = -\frac{1}{3}c_0, \qquad c_2 = \frac{1}{30}c_0, \qquad c_3 = -\frac{1}{630}c_0.$$

Two linearly independent solutions are

$$y_1 = 1 - x + \frac{1}{6}x^2 - \frac{1}{90}x^3 + \cdots$$

and

$$y_2 = x^{1/2}\left(1 - \frac{1}{3}x + \frac{1}{30}x^2 - \frac{1}{630}x^3 + \cdots\right).$$

12. Substituting $y = \sum_{n=0}^{\infty} c_n x^n$ into the DE we obtain

$$y'' - x^2 y' + xy = 2c_2 + (6c_3 + c_0)x + \sum_{k=1}^{\infty} [(k+3)(k+2)c_{k+3} - (k-1)c_k]x^{k+1} = 0$$

which implies $c_2 = 0$, $c_3 = -c_0/6$, and

$$c_{k+3} = \frac{k-1}{(k+3)(k+2)}c_k, \quad k = 1, 2, 3, \ldots .$$

Choosing $c_0 = 1$ and $c_1 = 0$ we find

$$c_3 = -\frac{1}{6}$$

$$c_4 = c_7 = c_{10} = \cdots = 0$$

$$c_5 = c_8 = c_{11} = \cdots = 0$$

$$c_6 = -\frac{1}{90}$$

and so on. For $c_0 = 0$ and $c_1 = 1$ we obtain

$$c_3 = c_6 = c_9 = \cdots = 0$$

$$c_4 = c_7 = c_{10} = \cdots = 0$$

$$c_5 = c_8 = c_{11} = \cdots = 0$$

and so on. Thus, two solutions are

$$y_1 = 1 - \frac{1}{6}x^3 - \frac{1}{90}x^6 - \cdots \quad \text{and} \quad y_2 = x.$$

15. Substituting $y = \sum_{n=0}^{\infty} c_n x^n$ into the DE we have

$$y'' + xy' + 2y = \underbrace{\sum_{n=2}^{\infty} n(n-1)c_n x^{n-2}}_{k=n-2} + \underbrace{\sum_{n=1}^{\infty} nc_n x^n}_{k=n} + 2\underbrace{\sum_{n=0}^{\infty} c_n x^n}_{k=n}$$

$$= \sum_{k=0}^{\infty} (k+2)(k+1)c_{k+2}x^k + \sum_{k=1}^{\infty} kc_k x^k + 2\sum_{k=0}^{\infty} c_k x^k$$

$$= 2c_2 + 2c_0 + \sum_{k=1}^{\infty} [(k+2)(k+1)c_{k+2} + (k+2)c_k]x^k = 0.$$

Thus

$$2c_2 + 2c_0 = 0$$

$$(k+2)(k+1)c_{k+2} + (k+2)c_k = 0$$

and

$$c_2 = -c_0$$

$$c_{k+2} = -\frac{1}{k+1} c_k, \quad k = 1, 2, 3, \ldots .$$

Choosing $c_0 = 1$ and $c_1 = 0$ we find

$$c_2 = -1$$

$$c_3 = c_5 = c_7 = \cdots = 0$$

$$c_4 = \frac{1}{3}$$

$$c_6 = -\frac{1}{15}$$

and so on. For $c_0 = 0$ and $c_1 = 1$ we obtain

$$c_2 = c_4 = c_6 = \cdots = 0$$

$$c_3 = -\frac{1}{2}$$

$$c_5 = \frac{1}{8}$$

$$c_7 = -\frac{1}{48}$$

and so on. Thus, the general solution is

$$y = C_0 \left(1 - x^2 + \frac{1}{3}x^4 - \frac{1}{15}x^6 + \cdots \right) + C_1 \left(x - \frac{1}{2}x^3 + \frac{1}{8}x^5 - \frac{1}{48}x^7 + \cdots \right)$$

and

$$y' = C_0 \left(-2x + \frac{4}{3}x^3 - \frac{2}{5}x^5 + \cdots \right) + C_1 \left(1 - \frac{3}{2}x^2 + \frac{5}{8}x^4 - \frac{7}{48}x^6 + \cdots \right).$$

Setting $y(0) = 3$ and $y'(0) = -2$ we find $c_0 = 3$ and $c_1 = -2$. Therefore, the solution of the initial-value problem is

$$y = 3 - 2x - 3x^2 + x^3 + x^4 - \frac{1}{4}x^5 - \frac{1}{5}x^6 + \frac{1}{24}x^7 + \cdots .$$

18. While we can find two solutions of the form

$$y_1 = c_0[1 + \cdots] \quad \text{and} \quad y_2 = c_1[x + \cdots],$$

the initial conditions at $x = 1$ give solutions for c_0 and c_1 in terms of infinite series. Letting $t = x - 1$ the initial-value problem becomes

$$\frac{d^2y}{dt^2} + (t+1)\frac{dy}{dt} + y = 0, \qquad y(0) = -6, \; y'(0) = 3.$$

Substituting $y = \sum_{n=0}^{\infty} c_n t^n$ into the DE, we have

$$\frac{d^2y}{dt^2} + (t+1)\frac{dy}{dt} + y = \underbrace{\sum_{n=2}^{\infty} n(n-1)c_n t^{n-2}}_{k=n-2} + \underbrace{\sum_{n=1}^{\infty} nc_n t^n}_{k=n} + \underbrace{\sum_{n=1}^{\infty} nc_n t^{n-1}}_{k=n-1} + \underbrace{\sum_{n=0}^{\infty} c_n t^n}_{k=n}$$

$$= \sum_{k=0}^{\infty}(k+2)(k+1)c_{k+2}t^k + \sum_{k=1}^{\infty} kc_k t^k + \sum_{k=0}^{\infty}(k+1)c_{k+1}t^k + \sum_{k=0}^{\infty} c_k t^k$$

$$= 2c_2 + c_1 + c_0 + \sum_{k=1}^{\infty}[(k+2)(k+1)c_{k+2} + (k+1)c_{k+1} + (k+1)c_k]t^k = 0.$$

Thus
$$2c_2 + c_1 + c_0 = 0$$

$$(k+2)(k+1)c_{k+2} + (k+1)c_{k+1} + (k+1)c_k = 0$$

and
$$c_2 = -\frac{c_1 + c_0}{2}$$

$$c_{k+2} = -\frac{c_{k+1} + c_k}{k+2}, \quad k = 1, 2, 3, \ldots.$$

Choosing $c_0 = 1$ and $c_1 = 0$ we find

$$c_2 = -\frac{1}{2}, \quad c_3 = \frac{1}{6}, \quad c_4 = \frac{1}{12},$$

and so on. For $c_0 = 0$ and $c_1 = 1$ we find

$$c_2 = -\frac{1}{2}, \quad c_3 = -\frac{1}{6}, \quad c_4 = \frac{1}{6},$$

and so on. Thus, the general solution is

$$y = c_0\left[1 - \frac{1}{2}t^2 + \frac{1}{6}t^3 + \frac{1}{12}t^4 + \cdots\right] + c_1\left[t - \frac{1}{2}t^2 - \frac{1}{6}t^3 + \frac{1}{6}t^4 + \cdots\right].$$

The initial conditions then imply $c_0 = -6$ and $c_1 = 3$. Thus the solution of the IVP is

$$y = -6\left[1 - \frac{1}{2}(x-1)^2 + \frac{1}{6}(x-1)^3 + \frac{1}{12}(x-1)^4 + \cdots\right]$$

$$+ 3\left[(x-1) - \frac{1}{2}(x-1)^2 - \frac{1}{6}(x-1)^3 + \frac{1}{6}(x-1)^4 + \cdots\right].$$

21. Substituting $y = \sum_{n=0}^{\infty} c_n x^n$ into the DE we have

$$y'' + x^2 y' + 2xy = \underbrace{\sum_{n=2}^{\infty} n(n-1)c_n x^{n-2}}_{k=n-2} + \underbrace{\sum_{n=1}^{\infty} nc_n x^{n+1}}_{k=n+1} + 2\underbrace{\sum_{n=0}^{\infty} c_n x^{n+1}}_{k=n+1}$$

$$= \sum_{k=0}^{\infty}(k+2)(k+1)c_{k+2}x^k + \sum_{k=2}^{\infty}(k-1)c_{k-1}x^k + 2\sum_{k=1}^{\infty} c_{k-1}x^k$$

$$= 2c_2 + (6c_3 + 2c_0)x + \sum_{k=2}^{\infty}[(k+2)(k+1)c_{k+2} + (k+1)c_{k-1}]x^k = 5 - 2x + 10x^3.$$

Thus, equating coefficients of like powers of x gives

$$2c_2 = 5$$

$$6c_3 + 2c_0 = -2$$

$$12c_4 + 3c_1 = 0$$

$$20c_5 + 4c_2 = 10$$

$$(k+2)(k+1)c_{k+2} + (k+1)c_{k-1} = 0, \quad k = 4, 5, 6, \ldots,$$

and

$$c_2 = \frac{5}{2}$$

$$c_3 = -\frac{1}{3}c_0 - \frac{1}{3}$$

$$c_4 = -\frac{1}{4}c_1$$

$$c_5 = \frac{1}{2} - \frac{1}{5}c_2 = \frac{1}{2} - \frac{1}{5}\left(\frac{5}{2}\right) = 0$$

$$c_{k+2} = -\frac{1}{k+2}c_{k-1}.$$

Using the recurrence relation, we find

$$c_6 = -\frac{1}{6}c_3 = \frac{1}{3 \cdot 6}(c_0 + 1) = \frac{1}{3^2 \cdot 2!}c_0 + \frac{1}{3^2 \cdot 2!}$$

$$c_7 = -\frac{1}{7}c_4 = \frac{1}{4 \cdot 7}c_1$$

$$c_8 = c_{11} = c_{14} = \cdots = 0$$

$$c_9 = -\frac{1}{9}c_6 = -\frac{1}{3^3 \cdot 3!}c_0 - \frac{1}{3^3 \cdot 3!}$$

$$c_{10} = -\frac{1}{10}c_7 = -\frac{1}{4 \cdot 7 \cdot 10}c_1$$

$$c_{12} = -\frac{1}{12}c_9 = \frac{1}{3^4 \cdot 4!}c_0 + \frac{1}{3^4 \cdot 4!}$$

$$c_{13} = -\frac{1}{13}c_0 = \frac{1}{4 \cdot 7 \cdot 10 \cdot 13}c_1,$$

and so on. Thus

$$y = c_0 \left[1 - \frac{1}{3}x^3 + \frac{1}{3^2 \cdot 2!}x^6 - \frac{1}{3^3 \cdot 3!}x^9 + \frac{1}{3^4 \cdot 4!}x^{12} - \cdots \right]$$

$$+ c_1 \left[x - \frac{1}{4}x^4 + \frac{1}{4 \cdot 7}x^7 - \frac{1}{4 \cdot 7 \cdot 10}x^{10} + \frac{1}{4 \cdot 7 \cdot 10 \cdot 13}x^{13} - \cdots \right]$$

$$+ \left[\frac{5}{2}x^2 - \frac{1}{3}x^3 + \frac{1}{3^2 \cdot 2!}x^6 - \frac{1}{3^3 \cdot 3!}x^9 + \frac{1}{3^4 \cdot 4!}x^{12} - \cdots \right].$$

24. **(a)** Using formula (5) of Section 4.2 in the text, we find that a second solution of $(1-x^2)y'' - 2xy' = 0$ is

$$y_2(x) = 1 \cdot \int \frac{e^{\int 2x\,dx/(1-x^2)}}{1^2}\,dx = \int e^{-\ln(1-x^2)}\,dx$$

$$= \int \frac{dx}{1-x^2} = \frac{1}{2}\ln\left(\frac{1+x}{1-x}\right),$$

where partial fractions was used to obtain the last integral.

(b) Using formula (5) of Section 4.2 in the text, we find that a second solution of $(1 - x^2)y'' - 2xy' + 2y = 0$ is

$$y_2(x) = x \cdot \int \frac{e^{\int 2x\,dx/(1-x^2)}}{x^2}\,dx = x \int \frac{e^{-\ln(1-x^2)}}{x^2}\,dx$$

$$= x \int \frac{dx}{x^2(1-x^2)}\,dx = x \left[\frac{1}{2}\ln\left(\frac{1+x}{1-x}\right) - \frac{1}{x} \right]$$

$$= \frac{x}{2}\ln\left(\frac{1+x}{1-x}\right) - 1,$$

where partial fractions was used to obtain the last integral.

(c)

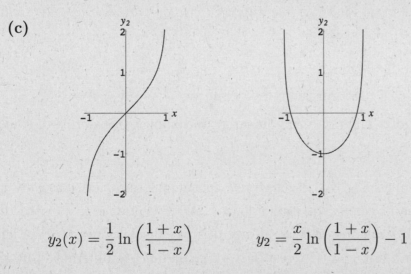

$$y_2(x) = \frac{1}{2}\ln\left(\frac{1+x}{1-x}\right) \qquad\qquad y_2 = \frac{x}{2}\ln\left(\frac{1+x}{1-x}\right) - 1$$

7 The Laplace Transform

Section 7.1

Definition of the Laplace Transform

The terminology and concepts listed below provide an outline of the main ideas encountered in this section. These can be useful when preparing for a quiz or test.

Terminology and Concepts

- integral transform
- definition of the Laplace transform
- linearity of the Laplace transform
- exponential order
- piecewise-continuous function
- transform of a piecewise-continuous function
- behavior of the Laplace transform $F(s)$ as $s \to \infty$

The basic skills listed below summarize the more mechanical types of problems encountered in the exercise set for this section.

Basic Skills

- find the Laplace transform of a function using the definition
- find the Laplace transform of a function using linearity and known transforms of some basic functions

Piecewise Continuity   A function f that is piecewise continuous on $[0, \infty)$ has only a finite number of *finite* discontinuities on any finite interval of $[0, \infty)$. Another name for a **finite discontinuity** is a **jump discontinuity**; this means that at any point of discontinuity t_1, the left- and right-hand limits, $\lim_{t \to t_1^-} f(t)$ and $\lim_{t \to t_1^+} f(t)$, respectively, exist but are different. Put another way, f cannot have any vertical asymptotes on the interval $[0, \infty)$.

Use of Computers   The Laplace transform function is built into version 5 of *Mathematica*. In *Maple* it is contained in the integral transforms package **inttrans**. The routines below define the Laplace transform of $f(t) = t^2 \cos t$ as a function $F(s)$.

```
Clear[f]                                              (Mathematica)
F[s_]:=Evaluate[LaplaceTransform[t^2 Cos[t], t, s]]
F[s]


with(inttrans);                                       (Maple)
F:=s->laplace(t^2*cos(t),t,s);
F(s);
```

Exercises 7.1   *Hints, Suggestions, Solutions, and Examples*

3. $\mathscr{L}\{f(t)\} = \displaystyle\int_0^1 te^{-st}dt + \int_1^\infty e^{-st}dt = \left(-\frac{1}{s}te^{-st} - \frac{1}{s^2}e^{-st}\right)\Big|_0^1 - \frac{1}{s}e^{-st}\Big|_1^\infty$

$\qquad = \left(-\dfrac{1}{s}e^{-s} - \dfrac{1}{s^2}e^{-s}\right) - \left(0 - \dfrac{1}{s^2}\right) - \dfrac{1}{s}(0 - e^{-s}) = \dfrac{1}{s^2}(1 - e^{-s}), \quad s > 0$

6. $\mathscr{L}\{f(t)\} = \displaystyle\int_{\pi/2}^\infty (\cos t)e^{-st}dt = \left(-\frac{s}{s^2+1}e^{-st}\cos t + \frac{1}{s^2+1}e^{-st}\sin t\right)\Big|_{\pi/2}^\infty$

$\qquad = 0 - \left(0 + \dfrac{1}{s^2+1}e^{-\pi s/2}\right) = -\dfrac{1}{s^2+1}e^{-\pi s/2}, \quad s > 0$

9. The function is $f(t) = \begin{cases} 1 - t, & 0 < t < 1 \\ 0, & t > 1 \end{cases}$ so

$\mathscr{L}\{f(t)\} = \displaystyle\int_0^1 (1-t)e^{-st}\,dt + \int_1^\infty 0e^{-st}\,dt = \int_0^1 (1-t)e^{-st}\,dt = \left(-\frac{1}{s}(1-t)e^{-st} + \frac{1}{s^2}e^{-st}\right)\Big|_0^1$

$\qquad = \dfrac{1}{s^2}e^{-s} + \dfrac{1}{s} - \dfrac{1}{s^2}, \quad s > 0$

12. $\mathscr{L}\{f(t)\} = \displaystyle\int_0^\infty e^{-2t-5}e^{-st}dt = e^{-5}\int_0^\infty e^{-(s+2)t}dt = -\frac{e^{-5}}{s+2}e^{-(s+2)t}\Big|_0^\infty = \frac{e^{-5}}{s+2}, \quad s > -2$

15. $\mathscr{L}\{f(t)\} = \displaystyle\int_0^\infty e^{-t}(\sin t)e^{-st}dt = \int_0^\infty (\sin t)e^{-(s+1)t}dt$

$\qquad = \left(\dfrac{-(s+1)}{(s+1)^2+1}e^{-(s+1)t}\sin t - \dfrac{1}{(s+1)^2+1}e^{-(s+1)t}\cos t\right)\Big|_0^\infty$

$\qquad = \dfrac{1}{(s+1)^2+1} = \dfrac{1}{s^2+2s+2}, \quad s > -1$

18. $\mathscr{L}\{f(t)\} = \int_0^\infty t(\sin t)e^{-st}\,dt$

$$= \left[\left(-\frac{t}{s^2+1} - \frac{2s}{(s^2+1)^2} \right)(\cos t)e^{-st} - \left(\frac{st}{s^2+1} + \frac{s^2-1}{(s^2+1)^2} \right)(\sin t)e^{-st} \right]_0^\infty$$

$$= \frac{2s}{(s^2+1)^2}, \quad s > 0$$

21. $\mathscr{L}\{4t - 10\} = \dfrac{4}{s^2} - \dfrac{10}{s}$

24. $\mathscr{L}\{-4t^2 + 16t + 9\} = -4\dfrac{2}{s^3} + \dfrac{16}{s^2} + \dfrac{9}{s}$

27. $\mathscr{L}\{1 + e^{4t}\} = \dfrac{1}{s} + \dfrac{1}{s-4}$

30. $\mathscr{L}\{e^{2t} - 2 + e^{-2t}\} = \dfrac{1}{s-2} - \dfrac{2}{s} + \dfrac{1}{s+2}$

33. $\mathscr{L}\{\sinh kt\} = \dfrac{1}{2}\mathscr{L}\{e^{kt} - e^{-kt}\} = \dfrac{1}{2}\left[\dfrac{1}{s-k} - \dfrac{1}{s+k} \right] = \dfrac{k}{s^2 - k^2}$

36. $\mathscr{L}\{e^{-t}\cosh t\} = \mathscr{L}\left\{ e^{-t}\dfrac{e^t + e^{-t}}{2} \right\} = \mathscr{L}\left\{ \dfrac{1}{2} + \dfrac{1}{2}e^{-2t} \right\} = \dfrac{1}{2s} + \dfrac{1}{2(s+2)}$

39. From the addition formula for the sine function, $\sin(4t + 5) = (\sin 4t)(\cos 5) + (\cos 4t)(\sin 5)$ so

$$\mathscr{L}\{\sin(4t + 5)\} = (\cos 5)\,\mathscr{L}\{\sin 4t\} + (\sin 5)\,\mathscr{L}\{\cos 4t\}$$

$$= (\cos 5)\frac{4}{s^2 + 16} + (\sin 5)\frac{s}{s^2 + 16}$$

$$= \frac{4\cos 5 + (\sin 5)s}{s^2 + 16}.$$

42. (a) $\mathscr{L}\{t^{-1/2}\} = \dfrac{\Gamma(1/2)}{s^{1/2}} = \sqrt{\dfrac{\pi}{s}}$

(b) $\mathscr{L}\{t^{1/2}\} = \dfrac{\Gamma(3/2)}{s^{3/2}} = \dfrac{\sqrt{\pi}}{2s^{3/2}}$

(c) $\mathscr{L}\{t^{3/2}\} = \dfrac{\Gamma(5/2)}{s^{5/2}} = \dfrac{3\sqrt{\pi}}{4s^{5/2}}$

43. (*Hint*) Think of a differentiable function that is of exponential order but whose derivative is not.

48. (*Hint*) Use integration by parts.

Section 7.2

Inverse Transforms and Transforms of Derivatives

The terminology and concepts listed below provide an outline of the main ideas encountered in this section. These can be useful when preparing for a quiz or test.

Terminology and Concepts

- definition of the inverse Laplace transform
- linearity of the inverse Laplace transform
- Laplace transform of the derivative of a function
- use of the Laplace transform to solve an IVP

The basic skills listed below summarize the more mechanical types of problems encountered in the exercise set for this section.

Basic Skills

- find the inverse Laplace transform of a function (frequently using partial fractions)
- solve a linear IVP using the Laplace transform

Partial Fractions Again   The key to success in working with the Laplace transform is the ability to do algebra quickly and accurately. Partial fraction decomposition is used almost every time you have to solve some sort of equation. In this manual, check out the review of partial fractions given in Section 2.2.

Use of Computers   As can be seen in this section, partial fractions are used extensively in the computation of inverse Laplace transforms. This process can be very computationally intensive, and as such, is ideally suited to a CAS. The examples below show how to find the partial fraction decomposition of $(s^2 + 1)/(s^3 + 4s)$.

```
Apart[(s^2+1)/(s^3+4s)]                          (Mathematica)
convert((s^2+1)/(s^3+4*s),parfrac,s);            (Maple)
```

The routines below define the inverse Laplace transform of $F(s) = (2s^3 - 6s)/(1 + s^2)^3$ as a function $f(t)$.

```
Clear[f]                                          (Mathematica)
f[t_]:=Evaluate[InverseLaplaceTransform[(2s^3 - 6s)/(s^2 + 1)^3, s, t]]
f[t]
```

```
with(inttrans);                                    (Maple)
f:=t->invlaplace((2*s^3-6*s)/(s^2+1)^3,s,t);
f(t);
```

Exercises 7.2   *Hints, Suggestions, Solutions, and Examples*

3. $\mathcal{L}^{-1}\left\{\dfrac{1}{s^2} - \dfrac{48}{s^5}\right\} = \mathcal{L}^{-1}\left\{\dfrac{1}{s^2} - \dfrac{48}{24}\cdot\dfrac{4!}{s^5}\right\} = t - 2t^4$

6. $\mathcal{L}^{-1}\left\{\dfrac{(s+2)^2}{s^3}\right\} = \mathcal{L}^{-1}\left\{\dfrac{1}{s} + 4\cdot\dfrac{1}{s^2} + 2\cdot\dfrac{2}{s^3}\right\} = 1 + 4t + 2t^2$

9. $\mathcal{L}^{-1}\left\{\dfrac{1}{4s+1}\right\} = \dfrac{1}{4}\mathcal{L}^{-1}\left\{\dfrac{1}{s+1/4}\right\} = \dfrac{1}{4}e^{-t/4}$

12. $\mathcal{L}^{-1}\left\{\dfrac{10s}{s^2+16}\right\} = 10\cos 4t$

15. $\mathcal{L}^{-1}\left\{\dfrac{2s-6}{s^2+9}\right\} = \mathcal{L}^{-1}\left\{2\cdot\dfrac{s}{s^2+9} - 2\cdot\dfrac{3}{s^2+9}\right\} = 2\cos 3t - 2\sin 3t$

18. $\mathcal{L}^{-1}\left\{\dfrac{s+1}{s^2-4s}\right\} = \mathcal{L}^{-1}\left\{-\dfrac{1}{4}\cdot\dfrac{1}{s} + \dfrac{5}{4}\cdot\dfrac{1}{s-4}\right\} = -\dfrac{1}{4} + \dfrac{5}{4}e^{4t}$

21. $\mathcal{L}^{-1}\left\{\dfrac{0.9s}{(s-0.1)(s+0.2)}\right\} = \mathcal{L}^{-1}\left\{(0.3)\cdot\dfrac{1}{s-0.1} + (0.6)\cdot\dfrac{1}{s+0.2}\right\} = 0.3e^{0.1t} + 0.6e^{-0.2t}$

24. $\mathcal{L}^{-1}\left\{\dfrac{s^2+1}{s(s-1)(s+1)(s-2)}\right\} = \mathcal{L}^{-1}\left\{\dfrac{1}{2}\cdot\dfrac{1}{s} - \dfrac{1}{s-1} - \dfrac{1}{3}\cdot\dfrac{1}{s+1} + \dfrac{5}{6}\cdot\dfrac{1}{s-2}\right\}$

$$= \dfrac{1}{2} - e^t - \dfrac{1}{3}e^{-t} + \dfrac{5}{6}e^{2t}$$

27. $\mathcal{L}^{-1}\left\{\dfrac{2s-4}{(s^2+s)(s^2+1)}\right\} = \mathcal{L}^{-1}\left\{\dfrac{2s-4}{s(s+1)(s^2+1)}\right\} = \mathcal{L}^{-1}\left\{-\dfrac{4}{s} + \dfrac{3}{s+1} + \dfrac{s}{s^2+1} + \dfrac{3}{s^2+1}\right\}$

$$= -4 + 3e^{-t} + \cos t + 3\sin t$$

30. $\mathcal{L}^{-1}\left\{\dfrac{6s+3}{(s^2+1)(s^2+4)}\right\} = \mathcal{L}^{-1}\left\{2\cdot\dfrac{s}{s^2+1} + \dfrac{1}{s^2+1} - 2\cdot\dfrac{s}{s^2+4} - \dfrac{1}{2}\cdot\dfrac{2}{s^2+4}\right\}$

$$= 2\cos t + \sin t - 2\cos 2t - \dfrac{1}{2}\sin 2t$$

33. The Laplace transform of the IVP is

$$s\mathcal{L}\{y\} - y(0) + 6\mathcal{L}\{y\} = \dfrac{1}{s-4}.$$

Solving for $\mathcal{L}\{y\}$ we obtain

$$\mathcal{L}\{y\} = \frac{1}{(s-4)(s+6)} + \frac{2}{s+6} = \frac{1}{10} \cdot \frac{1}{s-4} + \frac{19}{10} \cdot \frac{1}{s+6}.$$

Thus

$$y = \frac{1}{10}e^{4t} + \frac{19}{10}e^{-6t}.$$

36. The Laplace transform of the IVP is

$$s^2\mathcal{L}\{y\} - sy(0) - y'(0) - 4\left[s\mathcal{L}\{y\} - y(0)\right] = \frac{6}{s-3} - \frac{3}{s+1}.$$

Solving for $\mathcal{L}\{y\}$ we obtain

$$\mathcal{L}\{y\} = \frac{6}{(s-3)(s^2-4s)} - \frac{3}{(s+1)(s^2-4s)} + \frac{s-5}{s^2-4s}$$

$$= \frac{5}{2} \cdot \frac{1}{s} - \frac{2}{s-3} - \frac{3}{5} \cdot \frac{1}{s+1} + \frac{11}{10} \cdot \frac{1}{s-4}.$$

Thus

$$y = \frac{5}{2} - 2e^{3t} - \frac{3}{5}e^{-t} + \frac{11}{10}e^{4t}.$$

39. The Laplace transform of the IVP is

$$2\left[s^3\mathcal{L}\{y\} - s^2y(0) - sy'(0) - y''(0)\right] + 3[s^2\mathcal{L}\{y\} - sy(0) - y'(0)] - 3[s\mathcal{L}\{y\} - y(0)] - 2\mathcal{L}\{y\} = \frac{1}{s+1}.$$

Solving for $\mathcal{L}\{y\}$ we obtain

$$\mathcal{L}\{y\} = \frac{2s+3}{(s+1)(s-1)(2s+1)(s+2)} = \frac{1}{2}\frac{1}{s+1} + \frac{5}{18}\frac{1}{s-1} - \frac{8}{9}\frac{1}{s+1/2} + \frac{1}{9}\frac{1}{s+2}.$$

Thus

$$y = \frac{1}{2}e^{-t} + \frac{5}{18}e^t - \frac{8}{9}e^{-t/2} + \frac{1}{9}e^{-2t}.$$

42. The Laplace transform of the IVP is

$$s^2\mathcal{L}\{y\} - s \cdot 1 - 3 - 2[s\mathcal{L}\{y\} - 1] + 5\mathcal{L}\{y\} = (s^2 - 2s + 5)\mathcal{L}\{y\} - s - 1 = 0.$$

Solving for $\mathcal{L}\{y\}$ we obtain

$$\mathcal{L}\{y\} = \frac{s+1}{s^2 - 2s + 5} = \frac{s-1+2}{(s-1)^2 + 2^2} = \frac{s-1}{(s-1)^2 + 2^2} + \frac{2}{(s-1)^2 + 2^2}.$$

Thus

$$y = e^t \cos 2t + e^t \sin 2t.$$

43. (*Hint*) In part (b) compute $f''(t)$.

Section 7.3   **Operational Properties I**

The terminology and concepts listed below provide an outline of the main ideas encountered in this section. These can be useful when preparing for a quiz or test.

Terminology and Concepts

- first translation (or shifting) theorem
- unit step function (Heaviside function)
- second translation (or shifting) theorem

The basic skills listed below summarize the more mechanical types of problems encountered in the exercise set for this section.

Basic Skills

- use the first translation theorem to find the Laplace transform of a function that is a multiple of e^{at}.
- use the first translation theorem to find the inverse Laplace transform of the shifted function $F(s - a)$
- use the Laplace transform in conjunction with the first translation theorem to solve an IVP
- express a piecewise-defined function in terms of unit step functions
- find the Laplace transform of a piecewise-defined function
- find the inverse Laplace transform of a function that is a multiple of e^{-as}
- use the Laplace transform to solve an IVP involving piecewise-defined functions

Use of Computers   See Section 2.3 in this manual for *Mathematica* and *Maple* syntax for the unit step function.

Exercises 7.3   *Hints, Suggestions, Solutions, and Examples*

3. $\mathscr{L}\left\{t^3 e^{-2t}\right\} = \dfrac{3!}{(s+2)^4}$

6. $\mathscr{L}\left\{e^{2t}(t-1)^2\right\} = \mathscr{L}\left\{t^2 e^{2t} - 2te^{2t} + e^{2t}\right\} = \dfrac{2}{(s-2)^3} - \dfrac{2}{(s-2)^2} + \dfrac{1}{s-2}$

154

9. $\mathscr{L}\{(1 - e^t + 3e^{-4t})\cos 5t\} = \mathscr{L}\{\cos 5t - e^t \cos 5t + 3e^{-4t} \cos 5t\}$

$$= \frac{s}{s^2 + 25} - \frac{s - 1}{(s - 1)^2 + 25} + \frac{3(s + 4)}{(s + 4)^2 + 25}$$

12. $\mathscr{L}^{-1}\left\{\frac{1}{(s - 1)^4}\right\} = \frac{1}{6}\mathscr{L}^{-1}\left\{\frac{3!}{(s - 1)^4}\right\} = \frac{1}{6}t^3 e^t$

15. $\mathscr{L}^{-1}\left\{\frac{s}{s^2 + 4s + 5}\right\} = \mathscr{L}^{-1}\left\{\frac{s + 2}{(s + 2)^2 + 1^2} - 2\frac{1}{(s + 2)^2 + 1^2}\right\} = e^{-2t}\cos t - 2e^{-2t}\sin t$

18. $\mathscr{L}^{-1}\left\{\frac{5s}{(s - 2)^2}\right\} = \mathscr{L}^{-1}\left\{\frac{5(s - 2) + 10}{(s - 2)^2}\right\} = \mathscr{L}^{-1}\left\{\frac{5}{s - 2} + \frac{10}{(s - 2)^2}\right\} = 5e^{2t} + 10te^{2t}$

21. The Laplace transform of the DE is

$$s\mathscr{L}\{y\} - y(0) + 4\mathscr{L}\{y\} = \frac{1}{s + 4}.$$

Solving for $\mathscr{L}\{y\}$ we obtain

$$\mathscr{L}\{y\} = \frac{1}{(s + 4)^2} + \frac{2}{s + 4}.$$

Thus

$$y = te^{-4t} + 2e^{-4t}.$$

24. The Laplace transform of the DE is

$$s^2\mathscr{L}\{y\} - sy(0) - y'(0) - 4[s\mathscr{L}\{y\} - y(0)] + 4\mathscr{L}\{y\} = \frac{6}{(s - 2)^4}.$$

Solving for $\mathscr{L}\{y\}$ we obtain $\mathscr{L}\{y\} = \frac{1}{20}\frac{5!}{(s - 2)^6}$. Thus, $y = \frac{1}{20}t^5 e^{2t}$.

27. The Laplace transform of the DE is

$$s^2\mathscr{L}\{y\} - sy(0) - y'(0) - 6[s\mathscr{L}\{y\} - y(0)] + 13\mathscr{L}\{y\} = 0.$$

Solving for $\mathscr{L}\{y\}$ we obtain

$$\mathscr{L}\{y\} = -\frac{3}{s^2 - 6s + 13} = -\frac{3}{2}\frac{2}{(s - 3)^2 + 2^2}.$$

Thus

$$y = -\frac{3}{2}e^{3t}\sin 2t.$$

30. The Laplace transform of the DE is

$$s^2\mathscr{L}\{y\} - sy(0) - y'(0) - 2[s\mathscr{L}\{y\} - y(0)] + 5\mathscr{L}\{y\} = \frac{1}{s} + \frac{1}{s^2}.$$

Solving for $\mathcal{L}\{y\}$ we obtain

$$\mathcal{L}\{y\} = \frac{4s^2 + s + 1}{s^2(s^2 - 2s + 5)} = \frac{7}{25}\frac{1}{s} + \frac{1}{5}\frac{1}{s^2} + \frac{-7s/25 + 109/25}{s^2 - 2s + 5}$$

$$= \frac{7}{25}\frac{1}{s} + \frac{1}{5}\frac{1}{s^2} - \frac{7}{25}\frac{s-1}{(s-1)^2 + 2^2} + \frac{51}{25}\frac{2}{(s-1)^2 + 2^2}.$$

Thus

$$y = \frac{7}{25} + \frac{1}{5}t - \frac{7}{25}e^t\cos 2t + \frac{51}{25}e^t\sin 2t.$$

33. Recall from Section 5.1 that $mx'' = -kx - \beta x'$. Now $m = W/g = 4/32 = \frac{1}{8}$ slug, and $4 = 2k$ so that $k = 2$ lb/ft. Thus, the DE is $x'' + 7x' + 16x = 0$. The initial conditions are $x(0) = -3/2$ and $x'(0) = 0$. The Laplace transform of the DE is

$$s^2\mathcal{L}\{x\} + \frac{3}{2}s + 7s\mathcal{L}\{x\} + \frac{21}{2} + 16\mathcal{L}\{x\} = 0.$$

Solving for $\mathcal{L}\{x\}$ we obtain

$$\mathcal{L}\{x\} = \frac{-3s/2 - 21/2}{s^2 + 7s + 16} = -\frac{3}{2}\frac{s + 7/2}{(s + 7/2)^2 + (\sqrt{15}/2)^2} - \frac{7\sqrt{15}}{10}\frac{\sqrt{15}/2}{(s + 7/2)^2 + (\sqrt{15}/2)^2}.$$

Thus

$$x = -\frac{3}{2}e^{-7t/2}\cos\frac{\sqrt{15}}{2}t - \frac{7\sqrt{15}}{10}e^{-7t/2}\sin\frac{\sqrt{15}}{2}t.$$

36. The DE is

$$R\frac{dq}{dt} + \frac{1}{C}q = E_0 e^{-kt}, \quad q(0) = 0.$$

The Laplace transform of this equation is

$$Rs\,\mathcal{L}\{q\} + \frac{1}{C}\mathcal{L}\{q\} = E_0\frac{1}{s + k}.$$

Solving for $\mathcal{L}\{q\}$ we obtain

$$\mathcal{L}\{q\} = \frac{E_0 C}{(s + k)(RCs + 1)} = \frac{E_0/R}{(s + k)(s + 1/RC)}.$$

When $1/RC \neq k$ we have by partial fractions

$$\mathcal{L}\{q\} = \frac{E_0}{R}\left(\frac{1/(1/RC - k)}{s + k} - \frac{1/(1/RC - k)}{s + 1/RC}\right) = \frac{E_0}{R}\frac{1}{1/RC - k}\left(\frac{1}{s + k} - \frac{1}{s + 1/RC}\right).$$

Thus

$$q(t) = \frac{E_0 C}{1 - kRC}\left(e^{-kt} - e^{-t/RC}\right).$$

When $1/RC = k$ we have

$$\mathcal{L}\{q\} = \frac{E_0}{R}\frac{1}{(s + k)^2}.$$

Thus

$$q(t) = \frac{E_0}{R} t e^{-kt} = \frac{E_0}{R} t e^{-t/RC}.$$

39. $\mathscr{L}\{t\,\mathscr{U}(t-2)\} = \mathscr{L}\{(t-2)\,\mathscr{U}(t-2) + 2\,\mathscr{U}(t-2)\} = \dfrac{e^{-2s}}{s^2} + \dfrac{2e^{-2s}}{s}$

Alternatively, (16) of this section in the text could be used:

$$\mathscr{L}\{t\,\mathscr{U}(t-2)\} = e^{-2s}\,\mathscr{L}\{t+2\} = e^{-2s}\left(\frac{1}{s^2} + \frac{2}{s}\right).$$

42. $\mathscr{L}\left\{\sin t\,\mathscr{U}\left(t-\dfrac{\pi}{2}\right)\right\} = \mathscr{L}\left\{\cos\left(t-\dfrac{\pi}{2}\right)\mathscr{U}\left(t-\dfrac{\pi}{2}\right)\right\} = \dfrac{s e^{-\pi s/2}}{s^2+1}$

Alternatively, (16) of this section in the text could be used:

$$\mathscr{L}\left\{\sin t\,\mathscr{U}\left(t-\frac{\pi}{2}\right)\right\} = e^{-\pi s/2}\,\mathscr{L}\left\{\sin\left(t+\frac{\pi}{2}\right)\right\} = e^{-\pi s/2}\,\mathscr{L}\{\cos t\} = e^{-\pi s/2}\,\frac{s}{s^2+1}.$$

45. $\mathscr{L}^{-1}\left\{\dfrac{e^{-\pi s}}{s^2+1}\right\} = \sin(t-\pi)\,\mathscr{U}(t-\pi) = -\sin t\,\mathscr{U}(t-\pi)$

48. $\mathscr{L}^{-1}\left\{\dfrac{e^{-2s}}{s^2(s-1)}\right\} = \mathscr{L}^{-1}\left\{-\dfrac{e^{-2s}}{s} - \dfrac{e^{-2s}}{s^2} + \dfrac{e^{-2s}}{s-1}\right\} = -\mathscr{U}(t-2) - (t-2)\,\mathscr{U}(t-2) + e^{t-2}\,\mathscr{U}(t-2)$

51. The graph in Figure 7.3.13 is the portion of the graph in Figure 7.3.10 from $t=0$ to $t=b-a$ shifted a units to the right, with the rest of the graph zeroed out. This is the function in (f).

54. The graph in Figure 7.3.16 is the portion of the graph in Figure 7.3.10 from $t=0$ to $t=b$ with the rest of the graph zeroed out. This is the function in (d).

57. $\mathscr{L}\left\{t^2\,\mathscr{U}(t-1)\right\} = \mathscr{L}\left\{\left[(t-1)^2 + 2t - 1\right]\mathscr{U}(t-1)\right\} = \mathscr{L}\left\{\left[(t-1)^2 + 2(t-1) + 1\right]\mathscr{U}(t-1)\right\}$

$$= \left(\frac{2}{s^3} + \frac{2}{s^2} + \frac{1}{s}\right)e^{-s}$$

Alternatively, by (16) of this section in the text,

$$\mathscr{L}\{t^2\,\mathscr{U}(t-1)\} = e^{-s}\,\mathscr{L}\{t^2 + 2t + 1\} = e^{-s}\left(\frac{2}{s^3} + \frac{2}{s^2} + \frac{1}{s}\right).$$

60. $\mathscr{L}\{\sin t - \sin t\,\mathscr{U}(t-2\pi)\} = \mathscr{L}\{\sin t - \sin(t-2\pi)\,\mathscr{U}(t-2\pi)\} = \dfrac{1}{s^2+1} - \dfrac{e^{-2\pi s}}{s^2+1}$

63. The Laplace transform of the DE is

$$s\,\mathscr{L}\{y\} - y(0) + \mathscr{L}\{y\} = \frac{5}{s}e^{-s}.$$

Solving for $\mathscr{L}\{y\}$ we obtain

$$\mathscr{L}\{y\} = \frac{5e^{-s}}{s(s+1)} = 5e^{-s}\left[\frac{1}{s} - \frac{1}{s+1}\right].$$

Thus

$$y = 5\,\mathcal{U}(t-1) - 5e^{-(t-1)}\,\mathcal{U}(t-1).$$

66. The Laplace transform of the DE is

$$y'' + 4y = f(t) \qquad y(0)=0 \qquad y'(0)=-1$$

$$s^2 \mathcal{L}\{y\} - sy(0) - y'(0) + 4\mathcal{L}\{y\} = \frac{1}{s} - \frac{e^{-s}}{s}.$$

$$\begin{cases} 1 & 0 \le t < 1 \\ 0 & t \ge 1 \end{cases}$$

Solving for $\mathcal{L}\{y\}$ we obtain

$$\mathcal{L}\{y\} = \frac{1-s}{s(s^2+4)} - e^{-s}\frac{1}{s(s^2+4)} = \frac{1}{4}\frac{1}{s} - \frac{1}{4}\frac{s}{s^2+4} - \frac{1}{2}\frac{2}{s^2+4} - e^{-s}\left[\frac{1}{4}\frac{1}{s} - \frac{1}{4}\frac{s}{s^2+4}\right].$$

Thus

$$y = \frac{1}{4} - \frac{1}{4}\cos 2t - \frac{1}{2}\sin 2t - \left[\frac{1}{4} - \frac{1}{4}\cos 2(t-1)\right]\mathcal{U}(t-1).$$

69. The Laplace transform of the DE is

$$s^2 \mathcal{L}\{y\} - sy(0) - y'(0) + \mathcal{L}\{y\} = \frac{e^{-\pi s}}{s} - \frac{e^{-2\pi s}}{s}.$$

Solving for $\mathcal{L}\{y\}$ we obtain

$$\mathcal{L}\{y\} = e^{-\pi s}\left[\frac{1}{s} - \frac{s}{s^2+1}\right] - e^{-2\pi s}\left[\frac{1}{s} - \frac{s}{s^2+1}\right] + \frac{1}{s^2+1}.$$

Thus

$$y = [1 - \cos(t-\pi)]\,\mathcal{U}(t-\pi) - [1 - \cos(t-2\pi)]\,\mathcal{U}(t-2\pi) + \sin t.$$

72. Recall from Section 5.1 that $mx'' = -kx + f(t)$. Now $m = W/g = 32/32 = 1$ slug, and $32 = 2k$ so that $k = 16$ lb/ft. Thus, the differential equation is $x'' + 16x = f(t)$. The initial conditions are $x(0) = 0$, $x'(0) = 0$. Also, since

$$f(t) = \begin{cases} \sin t, & 0 \le t < 2\pi \\ 0, & t \ge 2\pi \end{cases}$$

and $\sin t = \sin(t - 2\pi)$ we can write

$$f(t) = \sin t - \sin(t - 2\pi)\,\mathcal{U}(t - 2\pi).$$

The Laplace transform of the DE is

$$s^2\mathcal{L}\{x\} + 16\mathcal{L}\{x\} = \frac{1}{s^2+1} - \frac{1}{s^2+1}e^{-2\pi s}.$$

Solving for $\mathcal{L}\{x\}$ we obtain

$$\mathcal{L}\{x\} = \frac{1}{(s^2+16)(s^2+1)} - \frac{1}{(s^2+16)(s^2+1)}e^{-2\pi s}$$

$$= \frac{-1/15}{s^2+16} + \frac{1/15}{s^2+1} - \left[\frac{-1/15}{s^2+16} + \frac{1/15}{s^2+1}\right]e^{-2\pi s}.$$

Thus

$$x(t) = -\frac{1}{60}\sin 4t + \frac{1}{15}\sin t + \frac{1}{60}\sin 4(t-2\pi)\mathscr{U}(t-2\pi) - \frac{1}{15}\sin(t-2\pi)\mathscr{U}(t-2\pi)$$

$$= \begin{cases} -\frac{1}{60}\sin 4t + \frac{1}{15}\sin t, & 0 \le t < 2\pi \\ 0, & t \ge 2\pi. \end{cases}$$

75. **(a)** The DE is

$$\frac{di}{dt} + 10i = \sin t + \cos\left(t - \frac{3\pi}{2}\right)\mathscr{U}\left(t - \frac{3\pi}{2}\right), \quad i(0) = 0.$$

The Laplace transform of this equation is

$$s\mathscr{L}\{i\} + 10\mathscr{L}\{i\} = \frac{1}{s^2+1} + \frac{se^{-3\pi s/2}}{s^2+1}.$$

Solving for $\mathscr{L}\{i\}$ we obtain

$$\mathscr{L}\{i\} = \frac{1}{(s^2+1)(s+10)} + \frac{s}{(s^2+1)(s+10)}e^{-3\pi s/2}$$

$$= \frac{1}{101}\left(\frac{1}{s+10} - \frac{s}{s^2+1} + \frac{10}{s^2+1}\right) + \frac{1}{101}\left(\frac{-10}{s+10} + \frac{10s}{s^2+1} + \frac{1}{s^2+1}\right)e^{-3\pi s/2},$$

Thus

$$i(t) = \frac{1}{101}\left(e^{-10t} - \cos t + 10\sin t\right)$$

$$+ \frac{1}{101}\left[-10e^{-10(t-3\pi/2)} + 10\cos\left(t - \frac{3\pi}{2}\right) + \sin\left(t - \frac{3\pi}{2}\right)\right]\mathscr{U}\left(t - \frac{3\pi}{2}\right).$$

(b)

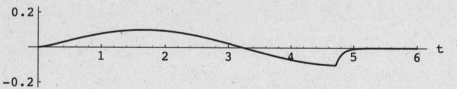

The maximum value of $i(t)$ is approximately 0.1 at $t = 1.7$, the minimum is approximately -0.1 at 4.7. [Using *Mathematica* we see that the maximum value is $i(t)$ is 0.0995037 at $t = 1.670465$, and the mininum value is $i(3\pi/2) \approx -0.0990099$ at $t = 3\pi/2$.]

78. The DE is

$$EI\frac{d^4y}{dx^4} = w_0[\mathscr{U}(x - L/3) - \mathscr{U}(x - 2L/3)].$$

Taking the Laplace transform of both sides and using $y(0) = y'(0) = 0$ we obtain

$$s^4\mathscr{L}\{y\} - sy''(0) - y'''(0) = \frac{w_0}{EI}\frac{1}{s}\left(e^{-Ls/3} - e^{-2Ls/3}\right).$$

Letting $y''(0) = c_1$ and $y'''(0) = c_2$ we have

$$\mathcal{L}\{y\} = \frac{c_1}{s^3} + \frac{c_2}{s^4} + \frac{w_0}{EI}\frac{1}{s^5}\left(e^{-Ls/3} - e^{-2Ls/3}\right)$$

so that

$$y(x) = \frac{1}{2}c_1x^2 + \frac{1}{6}c_2x^3 + \frac{1}{24}\frac{w_0}{EI}\left[\left(x - \frac{L}{3}\right)^4\mathcal{U}\left(x - \frac{L}{3}\right) - \left(x - \frac{2L}{3}\right)^4\mathcal{U}\left(x - \frac{2L}{3}\right)\right].$$

To find c_1 and c_2 we compute

$$y''(x) = c_1 + c_2x + \frac{1}{2}\frac{w_0}{EI}\left[\left(x - \frac{L}{3}\right)^2\mathcal{U}\left(x - \frac{L}{3}\right) - \left(x - \frac{2L}{3}\right)^2\mathcal{U}\left(x - \frac{2L}{3}\right)\right]$$

and

$$y'''(x) = c_2 + \frac{w_0}{EI}\left[\left(x - \frac{L}{3}\right)\mathcal{U}\left(x - \frac{L}{3}\right) - \left(x - \frac{2L}{3}\right)\mathcal{U}\left(x - \frac{2L}{3}\right)\right].$$

Then $y''(L) = y'''(L) = 0$ yields the system

$$c_1 + c_2L + \frac{1}{2}\frac{w_0}{EI}\left[\left(\frac{2L}{3}\right)^2 - \left(\frac{L}{3}\right)^2\right] = c_1 + c_2L + \frac{1}{6}\frac{w_0L^2}{EI} = 0$$

$$c_2 + \frac{w_0}{EI}\left[\frac{2L}{3} - \frac{L}{3}\right] = c_2 + \frac{1}{3}\frac{w_0L}{EI} = 0.$$

Solving for c_1 and c_2 we obtain $c_1 = \frac{1}{6}w_0L^2/EI$ and $c_2 = -\frac{1}{3}w_0L/EI$. Thus

$$y(x) = \frac{w_0}{EI}\left(\frac{1}{12}L^2x^2 - \frac{1}{18}Lx^3 + \frac{1}{24}\left[\left(x - \frac{L}{3}\right)^4\mathcal{U}\left(x - \frac{L}{3}\right) - \left(x - \frac{2L}{3}\right)^4\mathcal{U}\left(x - \frac{2L}{3}\right)\right]\right).$$

Section 7.4   Operational Properties II

The terminology and concepts listed below provide an outline of the main ideas encountered in this section. These can be useful when preparing for a quiz or test.

Terminology and Concepts

- the relationship between $t^n f(t)$ and the derivative of the Laplace transform of $f(t)$
- convolution of two functions
- the Laplace transform of the convolution f*g of two functions
- the Laplace transform of the integral of a function
- Volterra integral equation
- integrodifferential differential equation
- the Laplace transform of a periodic function

The basic skills listed below summarize the more mechanical types of problems encountered in the exercise set for this section.

Basic Skills

- find the Laplace transform of the product $t^n f(t)$ in terms of the Laplace transform of $f(t)$
- find the Laplace transform of the convolution of two functions
- find the Laplace transform of the integral of a function
- solve an integral equation
- solve an integrodifferential equation
- find the Laplace transform of a periodic function

DEs With Variable Coefficients

The Laplace transform is not well-suited to solving linear differential equations with variable coefficients. Nevertheless, it can be used in *some* instances. If the DE has coefficients that are at most linear polynomials in t, then its transform is a linear first-order DE in the transformed function $Y(s)$. For example, let's find a solution of $ty'' + 2ty' + 2y = t$ subject to $y(0) = 0$. From Theorem 7.4.1 observe

$$\mathscr{L}\{ty''\} = -\frac{d}{ds}\mathscr{L}\{y''\} = -\frac{d}{ds}[s^2Y(s) - sy(0) - y'(0)] = -s^2Y'(s) - 2sY(s) + y(0) \quad (1)$$

$$\mathscr{L}\{ty'\} = -\frac{d}{ds}\mathscr{L}\{y'\} = -\frac{d}{ds}[sY(s) - y(0)] = -sY'(s) - Y(s). \quad (2)$$

Note here that we have used the product rule to differentiate $s^2Y(s)$ and $sY(s)$. Hence the transform of the second-order DE is the first-order DE,

$$-s^2Y'(s) - 2sY(s) - 2sY'(s) - 2Y(s) + 2Y(s) = \frac{1}{s^2}$$

or

$$\frac{dY}{ds} + \frac{2}{s+2}Y = -\frac{1}{s^3(s+2)},$$

where we have used $y(0) = 0$. Since this last equation is recognized as a linear first-order DE in Y, we find from Section 2.3 in the text that the integrating factor is $(s+2)^2$. Multiplying the last equation by this factor gives

$$\frac{d}{ds}[(s+2)^2Y] = -\frac{s+2}{s^3} = -\frac{1}{s^2} - \frac{2}{s^3}.$$

Integrating, we obtain

$$(s+2)^2Y = \frac{1}{s} + \frac{1}{s^2} + c = \frac{s+1}{s^2} + c.$$

Now solve for Y and expand into partial fractions:

$$Y = \frac{s+1}{s^2(s+2)^2} + \frac{c}{(s+2)^2} = \frac{1}{4}\frac{1}{s^2} - \frac{1}{4}\frac{1}{(s+2)^2} + \frac{c}{(s+2)^2}.$$

The inverse of the preceding expression is then

$$y(t) = \frac{1}{4}t - \frac{1}{4}te^{-2t} + cte^{-2t} \qquad \text{or} \qquad y(t) = \frac{1}{4}t + c_1 te^{-2t},$$

where we have replaced the term $c - \frac{1}{4}$ by c_1.

Exercises 7.4   *Hints, Suggestions, Solutions, and Examples*

3. $\mathcal{L}\{t\cos 2t\} = -\dfrac{d}{ds}\left(\dfrac{s}{s^2+4}\right) = \dfrac{s^2-4}{(s^2+4)^2}$

6. $\mathcal{L}\{t^2\cos t\} = \dfrac{d^2}{ds^2}\left(\dfrac{s}{s^2+1}\right) = \dfrac{d}{ds}\left(\dfrac{1-s^2}{(s^2+1)^2}\right) = \dfrac{2s\left(s^2-3\right)}{(s^2+1)^3}$

9. The Laplace transform of the DE is

$$s\,\mathcal{L}\{y\} + \mathcal{L}\{y\} = \frac{2s}{(s^2+1)^2}.$$

Solving for $\mathcal{L}\{y\}$ we obtain

$$\mathcal{L}\{y\} = \frac{2s}{(s+1)(s^2+1)^2} = -\frac{1}{2}\frac{1}{s+1} - \frac{1}{2}\frac{1}{s^2+1} + \frac{1}{2}\frac{s}{s^2+1} + \frac{1}{(s^2+1)^2} + \frac{s}{(s^2+1)^2}.$$

Thus

$$y(t) = -\frac{1}{2}e^{-t} - \frac{1}{2}\sin t + \frac{1}{2}\cos t + \frac{1}{2}(\sin t - t\cos t) + \frac{1}{2}t\sin t$$

$$= -\frac{1}{2}e^{-t} + \frac{1}{2}\cos t - \frac{1}{2}t\cos t + \frac{1}{2}t\sin t.$$

12. The Laplace transform of the DE is

$$s^2\mathcal{L}\{y\} - sy(0) - y'(0) + \mathcal{L}\{y\} = \frac{1}{s^2+1}.$$

Solving for $\mathcal{L}\{y\}$ we obtain

$$\mathcal{L}\{y\} = \frac{s^3 - s^2 + s}{(s^2+1)^2} = \frac{s}{s^2+1} - \frac{1}{s^2+1} + \frac{1}{(s^2+1)^2}.$$

Thus

$$y = \cos t - \sin t + \left(\frac{1}{2}\sin t - \frac{1}{2}t\cos t\right) = \cos t - \frac{1}{2}\sin t - \frac{1}{2}t\cos t.$$

15.

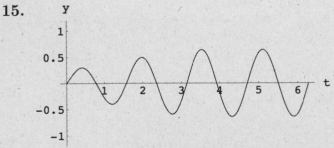

17. (*Hint*) See the discussion **DEs With Variable Coefficients** in this section for an example of how to solve this type of problem.

18. From Theorem 7.4.1 in the text

$$\mathscr{L}\{ty'\} = -\frac{d}{ds}\mathscr{L}\{y'\} = -\frac{d}{ds}[sY(s) - y(0)] = -s\frac{dY}{ds} - Y$$

so that the transform of the given second-order DE is the linear first-order DE in $Y(s)$:

$$Y' + \left(\frac{3}{s} - 2s\right)Y = -\frac{10}{s}.$$

Using the integrating factor $s^3 e^{-s^2}$, the last equation yields

$$Y(s) = \frac{5}{s^3} + \frac{c}{s^3}e^{s^2}.$$

But if $Y(s)$ is the Laplace transform of a piecewise-continuous function of exponential order, we must have, in view of Theorem 7.1.3, $\lim_{s\to\infty} Y(s) = 0$. In order to obtain this condition we require $c = 0$. Hence

$$y(t) = \mathscr{L}^{-1}\left\{\frac{5}{s^3}\right\} = \frac{5}{2}t^2.$$

21. $\mathscr{L}\{e^{-t} * e^t \cos t\} = \dfrac{s-1}{(s+1)\left[(s-1)^2 + 1\right]}$

24. $\mathscr{L}\left\{\displaystyle\int_0^t \cos\tau\,d\tau\right\} = \dfrac{1}{s}\mathscr{L}\{\cos t\} = \dfrac{s}{s(s^2+1)} = \dfrac{1}{s^2+1}$

27. $\mathscr{L}\left\{\displaystyle\int_0^t \tau e^{t-\tau}\,d\tau\right\} = \mathscr{L}\{t\}\mathscr{L}\{e^t\} = \dfrac{1}{s^2(s-1)}$

30. $\mathscr{L}\left\{t\displaystyle\int_0^t \tau e^{-\tau}d\tau\right\} = -\dfrac{d}{ds}\mathscr{L}\left\{\displaystyle\int_0^t \tau e^{-\tau}d\tau\right\} = -\dfrac{d}{ds}\left(\dfrac{1}{s}\dfrac{1}{(s+1)^2}\right) = \dfrac{3s+1}{s^2(s+1)^3}$

33. $\mathscr{L}^{-1}\left\{\dfrac{1}{s^3(s-1)}\right\} = \mathscr{L}^{-1}\left\{\dfrac{1/s^2(s-1)}{s}\right\} = \displaystyle\int_0^t (e^\tau - \tau - 1)d\tau = e^t - \dfrac{1}{2}t^2 - t - 1$

36. The Laplace transform of the DE is

$$s^2\mathscr{L}\{y\} + \mathscr{L}\{y\} = \frac{1}{(s^2+1)} + \frac{2s}{(s^2+1)^2}.$$

Thus

$$\mathscr{L}\{y\} = \frac{1}{(s^2+1)^2} + \frac{2s}{(s^2+1)^3}$$

and, using the result of Problem 35 with $k = 1$,

$$y = \frac{1}{2}(\sin t - t\cos t) + \frac{1}{4}(t\sin t - t^2\cos t).$$

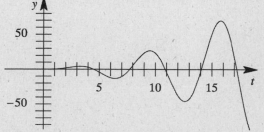

163

39. The Laplace transform of the given equation is

$$\mathscr{L}\{f\} = \mathscr{L}\{te^t\} + \mathscr{L}\{t\}\mathscr{L}\{f\}.$$

Solving for $\mathscr{L}\{f\}$ we obtain

$$\mathscr{L}\{f\} = \frac{s^2}{(s-1)^3(s+1)} = \frac{1}{8}\frac{1}{s-1} + \frac{3}{4}\frac{1}{(s-1)^2} + \frac{1}{4}\frac{2}{(s-1)^3} - \frac{1}{8}\frac{1}{s+1}.$$

Thus

$$f(t) = \frac{1}{8}e^t + \frac{3}{4}te^t + \frac{1}{4}t^2e^t - \frac{1}{8}e^{-t}.$$

42. The Laplace transform of the given equation is

$$\mathscr{L}\{f\} = \mathscr{L}\{\cos t\} + \mathscr{L}\{e^{-t}\}\mathscr{L}\{f\}.$$

Solving for $\mathscr{L}\{f\}$ we obtain

$$\mathscr{L}\{f\} = \frac{s}{s^2+1} + \frac{1}{s^2+1}.$$

Thus

$$f(t) = \cos t + \sin t.$$

45. The Laplace transform of the given equation is

$$s\mathscr{L}\{y\} - y(0) = \mathscr{L}\{1\} - \mathscr{L}\{\sin t\} - \mathscr{L}\{1\}\mathscr{L}\{y\}.$$

Solving for $\mathscr{L}\{f\}$ we obtain

$$\mathscr{L}\{y\} = \frac{s^2 - s + 1}{(s^2+1)^2} = \frac{1}{s^2+1} - \frac{1}{2}\frac{2s}{(s^2+1)^2}.$$

Thus

$$y = \sin t - \frac{1}{2}t\sin t.$$

48. The DE is

$$0.005\frac{di}{dt} + i + \frac{1}{0.02}\int_0^t i(\tau)d\tau = 100[t - (t-1)\,\mathscr{U}(t-1)]$$

or

$$\frac{di}{dt} + 200i + 10{,}000\int_0^t i(\tau)d\tau = 20{,}000[t - (t-1)\,\mathscr{U}(t-1)],$$

where $i(0) = 0$. The Laplace transform of the DE is

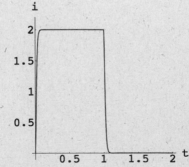

$$s\,\mathscr{L}\{i\} + 200\,\mathscr{L}\{i\} + \frac{10{,}000}{s}\mathscr{L}\{i\} = 20{,}000\left(\frac{1}{s^2} - \frac{1}{s^2}e^{-s}\right).$$

Solving for $\mathscr{L}\{i\}$ we obtain

$$\mathscr{L}\{i\} = \frac{20{,}000}{s(s+100)^2}(1 - e^{-s}) = \left[\frac{2}{s} - \frac{2}{s+100} - \frac{200}{(s+100)^2}\right](1 - e^{-s}).$$

Thus

$$i(t) = 2 - 2e^{-100t} - 200te^{-100t} - 2\mathscr{U}(t-1) + 2e^{-100(t-1)}\,\mathscr{U}(t-1) + 200(t-1)e^{-100(t-1)}\,\mathscr{U}(t-1).$$

51. Using integration by parts,

$$\mathscr{L}\{f(t)\} = \frac{1}{1 - e^{-bs}}\int_0^b \frac{a}{b}te^{-st}\,dt = \frac{a}{s}\left(\frac{1}{bs} - \frac{1}{e^{bs}-1}\right).$$

54. $\mathscr{L}\{f(t)\} = \dfrac{1}{1 - e^{-2\pi s}}\displaystyle\int_0^\pi e^{-st}\sin t\,dt = \dfrac{1}{s^2+1}\cdot\dfrac{1}{1 - e^{-\pi s}}$

57. The DE is $x'' + 2x' + 10x = 20f(t)$, where $f(t)$ is the meander function in Problem 49 with $a = \pi$. Using the initial conditions $x(0) = x'(0) = 0$ and taking the Laplace transform we obtain

$$(s^2 + 2s + 10)\,\mathscr{L}\{x(t)\} = \frac{20}{s}(1 - e^{-\pi s})\frac{1}{1 + e^{-\pi s}}$$

$$= \frac{20}{s}(1 - e^{-\pi s})(1 - e^{-\pi s} + e^{-2\pi s} - e^{-3\pi s} + \cdots)$$

$$= \frac{20}{s}(1 - 2e^{-\pi s} + 2e^{-2\pi s} - 2e^{-3\pi s} + \cdots)$$

$$= \frac{20}{s} + \frac{40}{s}\sum_{n=1}^{\infty}(-1)^n e^{-n\pi s}.$$

Then

$$\mathscr{L}\{x(t)\} = \frac{20}{s(s^2 + 2s + 10)} + \frac{40}{s(s^2 + 2s + 10)}\sum_{n=1}^{\infty}(-1)^n e^{-n\pi s}$$

$$= \frac{2}{s} - \frac{2s + 4}{s^2 + 2s + 10} + \sum_{n=1}^{\infty}(-1)^n\left[\frac{4}{s} - \frac{4s + 8}{s^2 + 2s + 10}\right]e^{-n\pi s}$$

$$= \frac{2}{s} - \frac{2(s+1) + 2}{(s+1)^2 + 9} + 4\sum_{n=1}^{\infty}(-1)^n\left[\frac{1}{s} - \frac{(s+1) + 1}{(s+1)^2 + 9}\right]e^{-n\pi s}$$

and

$$x(t) = 2\left(1 - e^{-t}\cos 3t - \frac{1}{3}e^{-t}\sin 3t\right) + 4\sum_{n=1}^{\infty}(-1)^n\left[1 - e^{-(t-n\pi)}\cos 3(t - n\pi)\right.$$

$$\left. - \frac{1}{3}e^{-(t-n\pi)}\sin 3(t - n\pi)\right]\mathscr{U}(t - n\pi).$$

The graph of $x(t)$ on the interval $[0, 2\pi)$ is shown below.

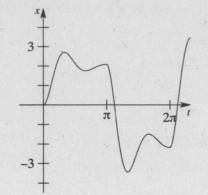

The Dirac Delta Function

The terminology and concepts listed below provide an outline of the main ideas encountered in this section. These can be useful when preparing for a quiz or test.

Terminology and Concepts

- unit impulse function
- Dirac delta function
- Laplace transform of the Dirac delta function

The basic skills listed below summarize the more mechanical types of problems encountered in the exercise set for this section.

Basic Skills

- solve a linear second-order DE whose input function involves the Dirac delta function

Exercises 7.5   *Hints, Suggestions, Solutions, and Examples*

In Problems 3, 6, and 9 we use the fact that $\sin t$ *is* 2π-*periodic, that is,* $\sin(t - 2\pi) = \sin t$.

3. The Laplace transform of the DE yields

$$\mathscr{L}\{y\} = \frac{1}{s^2 + 1}\left(1 + e^{-2\pi s}\right)$$

so that

$$y = \sin t + \sin t\, \mathcal{U}(t - 2\pi).$$

6. The Laplace transform of the DE yields

$$\mathcal{L}\{y\} = \frac{s}{s^2 + 1} + \frac{1}{s^2 + 1}(e^{-2\pi s} + e^{-4\pi s})$$

so that

$$y = \cos t + \sin t[\mathcal{U}(t - 2\pi) + \mathcal{U}(t - 4\pi)].$$

9. The Laplace transform of the DE yields

$$\mathcal{L}\{y\} = \frac{1}{(s + 2)^2 + 1}e^{-2\pi s}$$

so that

$$y = e^{-2(t - 2\pi)}\sin t\, \mathcal{U}(t - 2\pi).$$

12. The Laplace transform of the DE yields

$$\mathcal{L}\{y\} = \frac{1}{(s - 1)^2(s - 6)} + \frac{e^{-2s} + e^{-4s}}{(s - 1)(s - 6)}$$

$$= -\frac{1}{25}\frac{1}{s - 1} - \frac{1}{5}\frac{1}{(s - 1)^2} + \frac{1}{25}\frac{1}{s - 6} + \left[-\frac{1}{5}\frac{1}{s - 1} + \frac{1}{5}\frac{1}{s - 6}\right]\left(e^{-2s} + e^{-4s}\right)$$

so that

$$y = -\frac{1}{25}e^t - \frac{1}{5}te^t + \frac{1}{25}e^{6t} + \left[-\frac{1}{5}e^{t-2} + \frac{1}{5}e^{6(t-2)}\right]\mathcal{U}(t - 2)$$

$$+ \left[-\frac{1}{5}e^{t-4} + \frac{1}{5}e^{6(t-4)}\right]\mathcal{U}(t - 4).$$

15. (*Hint*) Consider the initial condition $y'(0) = 0$.

Section 7.6

Systems of Linear Differential Equations

The terminology and concepts listed below provide an outline of the main ideas encountered in this section. These can be useful when preparing for a quiz or test.

Terminology and Concepts

- coupled spring systems
- doubled pendulum

167

The basic skills listed below summarize the more mechanical types of problems encountered in the exercise set for this section.

Basic Skills

- use Laplace transform methods to solve a linear system of differential equations
- solve algebraic systems of equations by elimination

Exercises 7.6 *Hints, Suggestions, Solutions, and Examples*

3. Taking the Laplace transform of the system gives

$$s\mathcal{L}\{x\} + 1 = \mathcal{L}\{x\} - 2\mathcal{L}\{y\}$$

$$s\mathcal{L}\{y\} - 2 = 5\mathcal{L}\{x\} - \mathcal{L}\{y\}$$

so that

$$\mathcal{L}\{x\} = \frac{-s-5}{s^2+9} = -\frac{s}{s^2+9} - \frac{5}{3}\frac{3}{s^2+9}$$

and

$$x = -\cos 3t - \frac{5}{3}\sin 3t.$$

Then

$$y = \frac{1}{2}x - \frac{1}{2}x' = 2\cos 3t - \frac{7}{3}\sin 3t.$$

6. Taking the Laplace transform of the system gives

$$(s+1)\mathcal{L}\{x\} - (s-1)\mathcal{L}\{y\} = -1$$

$$s\mathcal{L}\{x\} + (s+2)\mathcal{L}\{y\} = 1$$

so that

$$\mathcal{L}\{y\} = \frac{s+1/2}{s^2+s+1} = \frac{s+1/2}{(s+1/2)^2 + (\sqrt{3}/2)^2}$$

and

$$\mathcal{L}\{x\} = \frac{-3/2}{s^2+s+1} = -\sqrt{3}\,\frac{\sqrt{3}/2}{(s+1/2)^2 + (\sqrt{3}/2)^2}.$$

Then

$$y = e^{-t/2}\cos\frac{\sqrt{3}}{2}t \quad \text{and} \quad x = -\sqrt{3}\,e^{-t/2}\sin\frac{\sqrt{3}}{2}t.$$

9. Adding the equations and then subtracting them gives

$$\frac{d^2x}{dt^2} = \frac{1}{2}t^2 + 2t$$

$$\frac{d^2y}{dt^2} = \frac{1}{2}t^2 - 2t.$$

168

Taking the Laplace transform of the system gives

$$\mathscr{L}\{x\} = 8\frac{1}{s} + \frac{1}{24}\frac{4!}{s^5} + \frac{1}{3}\frac{3!}{s^4}$$

and

$$\mathscr{L}\{y\} = \frac{1}{24}\frac{4!}{s^5} - \frac{1}{3}\frac{3!}{s^4}$$

so that

$$x = 8 + \frac{1}{24}t^4 + \frac{1}{3}t^3 \qquad \text{and} \qquad y = \frac{1}{24}t^4 - \frac{1}{3}t^3.$$

12. Taking the Laplace transform of the system gives

$$(s-4)\,\mathscr{L}\{x\} + 2\mathscr{L}\{y\} = \frac{2e^{-s}}{s}$$

$$-3\,\mathscr{L}\{x\} + (s+1)\,\mathscr{L}\{y\} = \frac{1}{2} + \frac{e^{-s}}{s}$$

so that

$$\mathscr{L}\{x\} = \frac{-1/2}{(s-1)(s-2)} + e^{-s}\frac{1}{(s-1)(s-2)}$$

$$= \frac{1}{2}\frac{1}{s-1} - \frac{1}{2}\frac{1}{s-2} + e^{-s}\left[-\frac{1}{s-1} + \frac{1}{s-2}\right]$$

and

$$\mathscr{L}\{y\} = \frac{e^{-s}}{s} + \frac{s/4 - 1}{(s-1)(s-2)} + e^{-s}\frac{-s/2 + 2}{(s-1)(s-2)}$$

$$= \frac{3}{4}\frac{1}{s-1} - \frac{1}{2}\frac{1}{s-2} + e^{-s}\left[\frac{1}{s} - \frac{3}{2}\frac{1}{s-1} + \frac{1}{s-2}\right].$$

Then

$$x = \frac{1}{2}e^t - \frac{1}{2}e^{2t} + \left[-e^{t-1} + e^{2(t-1)}\right]\mathscr{U}(t-1)$$

and

$$y = \frac{3}{4}e^t - \frac{1}{2}e^{2t} + \left[1 - \frac{3}{2}e^{t-1} + e^{2(t-1)}\right]\mathscr{U}(t-1).$$

15. (a) By Kirchhoff's first law we have $i_1 = i_2 + i_3$. By Kirchhoff's second law, on each loop we have $E(t) = Ri_1 + L_1 i_2'$ and $E(t) = Ri_1 + L_2 i_3'$ or $L_1 i_2' + Ri_2 + Ri_3 = E(t)$ and $L_2 i_3' + Ri_2 + Ri_3 = E(t)$.

(b) Taking the Laplace transform of the system

$$0.01i_2' + 5i_2 + 5i_3 = 100$$

$$0.0125i_3' + 5i_2 + 5i_3 = 100$$

gives

$$(s+500)\,\mathscr{L}\{i_2\} + 500\mathscr{L}\{i_3\} = \frac{10{,}000}{s}$$

$$400\mathscr{L}\{i_2\} + (s+400)\,\mathscr{L}\{i_3\} = \frac{8{,}000}{s}$$

so that

$$\mathscr{L}\{i_3\} = \frac{8,000}{s^2 + 900s} = \frac{80}{9}\frac{1}{s} - \frac{80}{9}\frac{1}{s+900}.$$

Then

$$i_3 = \frac{80}{9} - \frac{80}{9}e^{-900t} \quad \text{and} \quad i_2 = 20 - 0.0025i_3' - i_3 = \frac{100}{9} - \frac{100}{9}e^{-900t}.$$

(c) $i_1 = i_2 + i_3 = 20 - 20e^{-900t}$

18. Taking the Laplace transform of the system

$$0.5i_1' + 50i_2 = 60$$

$$0.005i_2' + i_2 - i_1 = 0$$

gives

$$s\,\mathscr{L}\{i_1\} + 100\,\mathscr{L}\{i_2\} = \frac{120}{s}$$

$$-200\,\mathscr{L}\{i_1\} + (s+200)\,\mathscr{L}\{i_2\} = 0$$

so that

$$\mathscr{L}\{i_2\} = \frac{24,000}{s(s^2 + 200s + 20,000)} = \frac{6}{5}\frac{1}{s} - \frac{6}{5}\frac{s+100}{(s+100)^2 + 100^2} - \frac{6}{5}\frac{100}{(s+100)^2 + 100^2}.$$

Then

$$i_2 = \frac{6}{5} - \frac{6}{5}e^{-100t}\cos 100t - \frac{6}{5}e^{-100t}\sin 100t$$

and

$$i_1 = 0.005i_2' + i_2 = \frac{6}{5} - \frac{6}{5}e^{-100t}\cos 100t.$$

Chapter 7 in Review

Hints, Suggestions, Solutions, and Examples

3. False; consider $f(t) = t^{-1/2}$.

6. False; consider $f(t) = 1$ and $g(t) = 1$.

9. $\mathscr{L}\{\sin 2t\} = \dfrac{2}{s^2 + 4}$

12. $\mathscr{L}\{\sin 2t\,\mathscr{U}(t - \pi)\} = \mathscr{L}\{\sin 2(t - \pi)\mathscr{U}(t - \pi)\} = \dfrac{2}{s^2 + 4}e^{-\pi s}$

15. $\mathcal{L}^{-1}\left\{\dfrac{1}{(s-5)^3}\right\} = \dfrac{1}{2}\,\mathcal{L}^{-1}\left\{\dfrac{2}{(s-5)^3}\right\} = \dfrac{1}{2}t^2 e^{5t}$

18. $\mathcal{L}^{-1}\left\{\dfrac{1}{s^2}e^{-5s}\right\} = (t-5)\,\mathcal{U}(t-5)$

21. $\mathcal{L}\left\{e^{-5t}\right\}$ exists for $s > -5$.

24. $\mathcal{L}\left\{\displaystyle\int_0^t e^{a\tau}f(\tau)\,d\tau\right\} = \dfrac{1}{s}\mathcal{L}\{e^{at}f(t)\} = \dfrac{F(s-a)}{s}$, whereas

$\mathcal{L}\left\{e^{at}\displaystyle\int_0^t f(\tau)\,d\tau\right\} = \mathcal{L}\left\{\displaystyle\int_0^t f(\tau)\,d\tau\right\}\bigg|_{s\to s-a} = \dfrac{F(s)}{s}\bigg|_{s\to s-a} = \dfrac{F(s-a)}{s-a}$.

27. $f(t-t_0)\,\mathcal{U}(t-t_0)$

30. $f(t) = \sin t\,\mathcal{U}(t-\pi) - \sin t\,\mathcal{U}(t-3\pi) = -\sin(t-\pi)\mathcal{U}(t-\pi) + \sin(t-3\pi)\mathcal{U}(t-3\pi)$

$\mathcal{L}\{f(t)\} = -\dfrac{1}{s^2+1}e^{-\pi s} + \dfrac{1}{s^2+1}e^{-3\pi s}$

$\mathcal{L}\{e^t f(t)\} = -\dfrac{1}{(s-1)^2+1}e^{-\pi(s-1)} + \dfrac{1}{(s-1)^2+1}e^{-3\pi(s-1)}$

33. Taking the Laplace transform of the DE we obtain

$$\mathcal{L}\{y\} = \dfrac{5}{(s-1)^2} + \dfrac{1}{2}\dfrac{2}{(s-1)^3}$$

so that

$$y = 5te^t + \dfrac{1}{2}t^2 e^t.$$

36. Taking the Laplace transform of the DE we obtain

$$\mathcal{L}\{y\} = \dfrac{s^3+2}{s^3(s-5)} - \dfrac{2+2s+s^2}{s^3(s-5)}e^{-s}$$

$$= -\dfrac{2}{125}\dfrac{1}{s} - \dfrac{2}{25}\dfrac{1}{s^2} - \dfrac{1}{5}\dfrac{2}{s^3} + \dfrac{127}{125}\dfrac{1}{s-5} - \left[-\dfrac{37}{125}\dfrac{1}{s} - \dfrac{12}{25}\dfrac{1}{s^2} - \dfrac{1}{5}\dfrac{2}{s^3} + \dfrac{37}{125}\dfrac{1}{s-5}\right]e^{-s}$$

so that

$$y = -\dfrac{2}{125} - \dfrac{2}{25}t - \dfrac{1}{5}t^2 + \dfrac{127}{125}e^{5t} - \left[-\dfrac{37}{125} - \dfrac{12}{25}(t-1) - \dfrac{1}{5}(t-1)^2 + \dfrac{37}{125}e^{5(t-1)}\right]\mathcal{U}(t-1).$$

39. Taking the Laplace transform of the system gives

$$s\mathcal{L}\{x\} + \mathcal{L}\{y\} = \dfrac{1}{s^2} + 1$$

$$4\mathcal{L}\{x\} + s\mathcal{L}\{y\} = 2$$

so that

$$\mathscr{L}\{x\} = \frac{s^2 - 2s + 1}{s(s-2)(s+2)} = -\frac{1}{4}\frac{1}{s} + \frac{1}{8}\frac{1}{s-2} + \frac{9}{8}\frac{1}{s+2}.$$

Then

$$x = -\frac{1}{4} + \frac{1}{8}e^{2t} + \frac{9}{8}e^{-2t} \quad \text{and} \quad y = -x' + t = \frac{9}{4}e^{-2t} - \frac{1}{4}e^{2t} + t.$$

42. The DE is

$$\frac{1}{2}\frac{d^2q}{dt^2} + 10\frac{dq}{dt} + 100q = 10 - 10\,\mathscr{U}(t-5).$$

Taking the Laplace transform we obtain

$$\mathscr{L}\{q\} = \frac{20}{s(s^2 + 20s + 200)}\left(1 - e^{-5s}\right)$$

$$= \left[\frac{1}{10}\frac{1}{s} - \frac{1}{10}\frac{s+10}{(s+10)^2 + 10^2} - \frac{1}{10}\frac{10}{(s+10)^2 + 10^2}\right]\left(1 - e^{-5s}\right)$$

so that

$$q(t) = \frac{1}{10} - \frac{1}{10}e^{-10t}\cos 10t - \frac{1}{10}e^{-10t}\sin 10t$$

$$- \left[\frac{1}{10} - \frac{1}{10}e^{-10(t-5)}\cos 10(t-5) - \frac{1}{10}e^{-10(t-5)}\sin 10(t-5)\right]\mathscr{U}(t-5).$$

45. (a) With $\omega^2 = g/l$ and $K = k/m$ the system of DEs is

$$\theta_1'' + \omega^2\theta_1 = -K(\theta_1 - \theta_2)$$

$$\theta_2'' + \omega^2\theta_2 = K(\theta_1 - \theta_2).$$

Denoting the Laplace transform of $\theta(t)$ by $\Theta(s)$ we have that the Laplace transform of the system is

$$(s^2 + \omega^2)\Theta_1(s) = -K\Theta_1(s) + K\Theta_2(s) + s\theta_0$$

$$(s^2 + \omega^2)\Theta_2(s) = K\Theta_1(s) - K\Theta_2(s) + s\psi_0.$$

If we add the two equations, we get

$$\Theta_1(s) + \Theta_2(s) = (\theta_0 + \psi_0)\frac{s}{s^2 + \omega^2}$$

which implies

$$\theta_1(t) + \theta_2(t) = (\theta_0 + \psi_0)\cos\omega t.$$

This enables us to solve for first, say, $\theta_1(t)$ and then find $\theta_2(t)$ from

$$\theta_2(t) = -\theta_1(t) + (\theta_0 + \psi_0)\cos\omega t.$$

Now solving

$$(s^2 + \omega^2 + K)\Theta_1(s) - K\Theta_2(s) = s\theta_0$$

$$-k\Theta_1(s) + (s^2 + \omega^2 + K)\Theta_2(s) = s\psi_0$$

gives

$$[(s^2 + \omega^2 + K)^2 - K^2]\Theta_1(s) = s(s^2 + \omega^2 + K)\theta_0 + Ks\psi_0.$$

Factoring the difference of two squares and using partial fractions we get

$$\Theta_1(s) = \frac{s(s^2 + \omega^2 + K)\theta_0 + Ks\psi_0}{(s^2 + \omega^2)(s^2 + \omega^2 + 2K)} = \frac{\theta_0 + \psi_0}{2} \frac{s}{s^2 + \omega^2} + \frac{\theta_0 - \psi_0}{2} \frac{s}{s^2 + \omega^2 + 2K},$$

so

$$\theta_1(t) = \frac{\theta_0 + \psi_0}{2} \cos \omega t + \frac{\theta_0 - \psi_0}{2} \cos \sqrt{\omega^2 + 2K}\, t.$$

Then from $\theta_2(t) = -\theta_1(t) + (\theta_0 + \psi_0)\cos \omega t$ we get

$$\theta_2(t) = \frac{\theta_0 + \psi_0}{2} \cos \omega t - \frac{\theta_0 - \psi_0}{2} \cos \sqrt{\omega^2 + 2K}\, t.$$

(b) With the initial conditions $\theta_1(0) = \theta_0$, $\theta_1'(0) = 0$, $\theta_2(0) = \theta_0$, $\theta_2'(0) = 0$ we have

$$\theta_1(t) = \theta_0 \cos \omega t, \qquad \theta_2(t) = \theta_0 \cos \omega t.$$

Physically this means that both pendulums swing in the same direction as if they were free since the spring exerts no influence on the motion ($\theta_1(t)$ and $\theta_2(t)$ are free of K).

With the initial conditions $\theta_1(0) = \theta_0$, $\theta_1'(0) = 0$, $\theta_2(0) = -\theta_0$, $\theta_2'(0) = 0$ we have

$$\theta_1(t) = \theta_0 \cos \sqrt{\omega^2 + 2K}\, t, \qquad \theta_2(t) = -\theta_0 \cos \sqrt{\omega^2 + 2K}\, t.$$

Physically this means that both pendulums swing in the opposite directions, stretching and compressing the spring. The amplitude of both displacements is $|\theta_0|$. Moreover, $\theta_1(t) = \theta_0$ and $\theta_2(t) = -\theta_0$ at precisely the same times. At these times the spring is stretched to its maximum.

8 Systems of Linear First-Order Differential Equations

Preliminary Theory—Linear Systems

The terminology and concepts listed below provide an outline of the main ideas encountered in this section. These can be useful when preparing for a quiz or test.

Terminology and Concepts

- linear system of first-order DEs

- homogeneous linear system of DEs

- matrix form of a linear system of DEs

- superposition principle

- linear independence of a set of solutions

- Wronskian

- fundamental set of solutions

- general solution of a linear system of DEs

- particular solution of a nonhomogeneous linear system of DEs

- complementary function of a homogeneous linear system of DEs

The basic skills listed below summarize the more mechanical types of problems encountered in the exercise set for this section.

Basic Skills

- express a linear first-order system in both normal form and matrix form

- determine whether a set of vectors forms a fundamental set

3. Let $\mathbf{X} = \begin{pmatrix} x \\ y \\ z \end{pmatrix}$. Then $\mathbf{X}' = \begin{pmatrix} -3 & 4 & -9 \\ 6 & -1 & 0 \\ 10 & 4 & 3 \end{pmatrix} \mathbf{X}$.

6. Let $\mathbf{X} = \begin{pmatrix} x \\ y \\ z \end{pmatrix}$. Then $\mathbf{X}' = \begin{pmatrix} -3 & 4 & 0 \\ 5 & 9 & 0 \\ 0 & 1 & 6 \end{pmatrix} \mathbf{X} + \begin{pmatrix} e^{-t}\sin 2t \\ 4e^{-t}\cos 2t \\ -e^{-t} \end{pmatrix}$.

9. $\dfrac{dx}{dt} = x - y + 2z + e^{-t} - 3t$

$\dfrac{dy}{dt} = 3x - 4y + z + 2e^{-t} + t$

$\dfrac{dz}{dt} = -2x + 5y + 6z + 2e^{-t} - t$

12. Since

$$\mathbf{X}' = \begin{pmatrix} 5\cos t - 5\sin t \\ 2\cos t - 4\sin t \end{pmatrix} e^t \quad \text{and} \quad \begin{pmatrix} -2 & 5 \\ -2 & 4 \end{pmatrix} \mathbf{X} = \begin{pmatrix} 5\cos t - 5\sin t \\ 2\cos t - 4\sin t \end{pmatrix} e^t$$

we see that

$$\mathbf{X}' = \begin{pmatrix} -2 & 5 \\ -2 & 4 \end{pmatrix} \mathbf{X}.$$

15. Since

$$\mathbf{X}' = \begin{pmatrix} 0 \\ 0 \\ 0 \end{pmatrix} \quad \text{and} \quad \begin{pmatrix} 1 & 2 & 1 \\ 6 & -1 & 0 \\ -1 & -2 & -1 \end{pmatrix} \mathbf{X} = \begin{pmatrix} 0 \\ 0 \\ 0 \end{pmatrix}$$

we see that

$$\mathbf{X}' = \begin{pmatrix} 1 & 2 & 1 \\ 6 & -1 & 0 \\ -1 & -2 & -1 \end{pmatrix} \mathbf{X}.$$

18. Yes, since $W(\mathbf{X}_1, \mathbf{X}_2) = 8e^{2t} \neq 0$ the set $\mathbf{X}_1, \mathbf{X}_2$ is linearly independent on $-\infty < t < \infty$.

21. Since

$$\mathbf{X}'_p = \begin{pmatrix} 2 \\ -1 \end{pmatrix} \quad \text{and} \quad \begin{pmatrix} 1 & 4 \\ 3 & 2 \end{pmatrix} \mathbf{X}_p + \begin{pmatrix} 2 \\ -4 \end{pmatrix} t + \begin{pmatrix} -7 \\ -18 \end{pmatrix} = \begin{pmatrix} 2 \\ -1 \end{pmatrix}$$

we see that

$$\mathbf{X}'_p = \begin{pmatrix} 1 & 4 \\ 3 & 2 \end{pmatrix} \mathbf{X}_p + \begin{pmatrix} 2 \\ -4 \end{pmatrix} t + \begin{pmatrix} -7 \\ -18 \end{pmatrix}.$$

24. Since

$$\mathbf{X}_p' = \begin{pmatrix} 3\cos 3t \\ 0 \\ -3\sin 3t \end{pmatrix} \quad \text{and} \quad \begin{pmatrix} 1 & 2 & 3 \\ -4 & 2 & 0 \\ -6 & 1 & 0 \end{pmatrix} \mathbf{X}_p + \begin{pmatrix} -1 \\ 4 \\ 3 \end{pmatrix} \sin 3t = \begin{pmatrix} 3\cos 3t \\ 0 \\ -3\sin 3t \end{pmatrix}$$

we see that

$$\mathbf{X}_p' = \begin{pmatrix} 1 & 2 & 3 \\ -4 & 2 & 0 \\ -6 & 1 & 0 \end{pmatrix} \mathbf{X}_p + \begin{pmatrix} -1 \\ 4 \\ 3 \end{pmatrix} \sin 3t.$$

Section 8.2 — Homogeneous Linear Systems

The terminology and concepts listed below provide an outline of the main ideas encountered in this section. These can be useful when preparing for a quiz or test.

Terminology and Concepts

- characteristic equation of a matrix
- eigenvalues of a matrix
- eigenvector of a matrix corresponding to an eigenvalue
- general solution of a homogeneous linear system of DEs with constant coefficients
- phase plane
- trajectory
- phase portrait of a linear system of DEs
- repeller
- attractor
- multiplicity of an eigenvalue

The basic skills listed below summarize the more mechanical types of problems encountered in the exercise set for this section.

Basic Skills

- find the general solution of a homogeneous linear system of DEs with constant coefficients when the eigenvalues are real and distinct, real and repeated, or complex

Use of Computers To find eigenvalues and eigenvectors of a matrix using *Mathematica* or *Maple*, we can input the matrix by rows. For example,

$$m = \begin{pmatrix} -4 & 1 & 1 \\ 1 & 5 & -1 \\ 0 & 1 & -3 \end{pmatrix}$$

is the matrix of the coefficients of the linear system (6) in Example 2 in the text. Using *Mathematica*, we can write **m** as

m={{-4, 1, 1}, {1, 5, -1}, {0, 1, -3}}

The commands **Eigenvalues[m]** and **Eigenvectors[m]** given in sequence yield

$$[-4, -3, 5] \quad \text{and} \quad \{\{-10, -1, 1\}, \{1, 0, 1\}, \{1, 8, 1\}\}.$$

Translated this means $\lambda = -4$ is an eigenvalue with corresponding eigenvector

$$\mathbf{K} = \begin{pmatrix} -10 \\ -1 \\ 1 \end{pmatrix},$$

and so on. In *Mathematica* eigenvalues and eigenvectors can be obtained at the same time using the single command **Eigensystem[m]**. In *Maple*, the matrix **m** is written

m:=matrix(3,3,[-4,1,1,1,5,-1,0,1,-3])

The commands for eigenvalues and eigenvectors are, respectively, **eigenvals(m)** and **eigenvects(m)**.

Exercises 8.2   *Hints, Suggestions, Solutions, and Examples*

3. The system is

$$\mathbf{X}' = \begin{pmatrix} -4 & 2 \\ -5/2 & 2 \end{pmatrix} \mathbf{X}$$

and $\det(\mathbf{A} - \lambda\mathbf{I}) = (\lambda - 1)(\lambda + 3) = 0$. For $\lambda_1 = 1$ we obtain

$$\begin{pmatrix} -5 & 2 & | & 0 \\ -5/2 & 1 & | & 0 \end{pmatrix} \Longrightarrow \begin{pmatrix} -5 & 2 & | & 0 \\ 0 & 0 & | & 0 \end{pmatrix} \quad \text{so that} \quad \mathbf{K}_1 = \begin{pmatrix} 2 \\ 5 \end{pmatrix}.$$

For $\lambda_2 = -3$ we obtain

$$\begin{pmatrix} -1 & 2 & | & 0 \\ -5/2 & 5 & | & 0 \end{pmatrix} \Longrightarrow \begin{pmatrix} -1 & 2 & | & 0 \\ 0 & 0 & | & 0 \end{pmatrix} \quad \text{so that} \quad \mathbf{K}_2 = \begin{pmatrix} 2 \\ 1 \end{pmatrix}.$$

Then

$$\mathbf{X} = c_1 \begin{pmatrix} 2 \\ 5 \end{pmatrix} e^t + c_2 \begin{pmatrix} 2 \\ 1 \end{pmatrix} e^{-3t}.$$

6. The system is

$$\mathbf{X}' = \begin{pmatrix} -6 & 2 \\ -3 & 1 \end{pmatrix} \mathbf{X}$$

and $\det(\mathbf{A} - \lambda\mathbf{I}) = \lambda(\lambda + 5) = 0$. For $\lambda_1 = 0$ we obtain

$$\begin{pmatrix} -6 & 2 & \Big| & 0 \\ -3 & 1 & \Big| & 0 \end{pmatrix} \implies \begin{pmatrix} 1 & -1/3 & \Big| & 0 \\ 0 & 0 & \Big| & 0 \end{pmatrix} \quad \text{so that} \quad \mathbf{K}_1 = \begin{pmatrix} 1 \\ 3 \end{pmatrix}.$$

For $\lambda_2 = -5$ we obtain

$$\begin{pmatrix} -1 & 2 & \Big| & 0 \\ -3 & 6 & \Big| & 0 \end{pmatrix} \implies \begin{pmatrix} 1 & -2 & \Big| & 0 \\ 0 & 0 & \Big| & 0 \end{pmatrix} \quad \text{so that} \quad \mathbf{K}_2 = \begin{pmatrix} 2 \\ 1 \end{pmatrix}.$$

Then

$$\mathbf{X} = c_1 \begin{pmatrix} 1 \\ 3 \end{pmatrix} + c_2 \begin{pmatrix} 2 \\ 1 \end{pmatrix} e^{-5t}.$$

9. We have $\det(\mathbf{A} - \lambda\mathbf{I}) = -(\lambda + 1)(\lambda - 3)(\lambda + 2) = 0$. For $\lambda_1 = -1$, $\lambda_2 = 3$, and $\lambda_3 = -2$ we obtain

$$\mathbf{K}_1 = \begin{pmatrix} -1 \\ 0 \\ 1 \end{pmatrix}, \quad \mathbf{K}_2 = \begin{pmatrix} 1 \\ 4 \\ 3 \end{pmatrix}, \quad \text{and} \quad \mathbf{K}_3 = \begin{pmatrix} 1 \\ -1 \\ 3 \end{pmatrix},$$

so that

$$\mathbf{X} = c_1 \begin{pmatrix} -1 \\ 0 \\ 1 \end{pmatrix} e^{-t} + c_2 \begin{pmatrix} 1 \\ 4 \\ 3 \end{pmatrix} e^{3t} + c_3 \begin{pmatrix} 1 \\ -1 \\ 3 \end{pmatrix} e^{-2t}.$$

12. We have $\det(\mathbf{A} - \lambda\mathbf{I}) = (\lambda - 3)(\lambda + 5)(6 - \lambda) = 0$. For $\lambda_1 = 3$, $\lambda_2 = -5$, and $\lambda_3 = 6$ we obtain

$$\mathbf{K}_1 = \begin{pmatrix} 1 \\ 1 \\ 0 \end{pmatrix}, \quad \mathbf{K}_2 = \begin{pmatrix} 1 \\ -1 \\ 0 \end{pmatrix}, \quad \text{and} \quad \mathbf{K}_3 = \begin{pmatrix} 2 \\ -2 \\ 11 \end{pmatrix},$$

so that

$$\mathbf{X} = c_1 \begin{pmatrix} 1 \\ 1 \\ 0 \end{pmatrix} e^{3t} + c_2 \begin{pmatrix} 1 \\ -1 \\ 0 \end{pmatrix} e^{-5t} + c_3 \begin{pmatrix} 2 \\ -2 \\ 11 \end{pmatrix} e^{6t}.$$

21. We have $\det(\mathbf{A} - \lambda\mathbf{I}) = (\lambda - 2)^2 = 0$. For $\lambda_1 = 2$ we obtain

$$\mathbf{K} = \begin{pmatrix} 1 \\ 1 \end{pmatrix}.$$

A solution of $(\mathbf{A} - \lambda_1\mathbf{I})\mathbf{P} = \mathbf{K}$ is

$$\mathbf{P} = \begin{pmatrix} -1/3 \\ 0 \end{pmatrix}$$

so that

$$\mathbf{X} = c_1 \begin{pmatrix} 1 \\ 1 \end{pmatrix} e^{2t} + c_2 \left[\begin{pmatrix} 1 \\ 1 \end{pmatrix} te^{2t} + \begin{pmatrix} -1/3 \\ 0 \end{pmatrix} e^{2t} \right].$$

24. We have $\det(\mathbf{A} - \lambda\mathbf{I}) = (\lambda - 8)(\lambda + 1)^2 = 0$. For $\lambda_1 = 8$ we obtain

$$\mathbf{K}_1 = \begin{pmatrix} 2 \\ 1 \\ 2 \end{pmatrix}.$$

For $\lambda_2 = -1$ we obtain

$$\mathbf{K}_2 = \begin{pmatrix} 0 \\ -2 \\ 1 \end{pmatrix} \quad \text{and} \quad \mathbf{K}_3 = \begin{pmatrix} 1 \\ -2 \\ 0 \end{pmatrix}.$$

Then

$$\mathbf{X} = c_1 \begin{pmatrix} 2 \\ 1 \\ 2 \end{pmatrix} e^{8t} + c_2 \begin{pmatrix} 0 \\ -2 \\ 1 \end{pmatrix} e^{-t} + c_3 \begin{pmatrix} 1 \\ -2 \\ 0 \end{pmatrix} e^{-t}.$$

27. We have $\det(\mathbf{A} - \lambda\mathbf{I}) = -(\lambda - 1)^3 = 0$. For $\lambda_1 = 1$ we obtain

$$\mathbf{K} = \begin{pmatrix} 0 \\ 1 \\ 1 \end{pmatrix}.$$

Solutions of $(\mathbf{A} - \lambda_1\mathbf{I})\mathbf{P} = \mathbf{K}$ and $(\mathbf{A} - \lambda_1\mathbf{I})\mathbf{Q} = \mathbf{P}$ are

$$\mathbf{P} = \begin{pmatrix} 0 \\ 1 \\ 0 \end{pmatrix} \quad \text{and} \quad \mathbf{Q} = \begin{pmatrix} 1/2 \\ 0 \\ 0 \end{pmatrix}$$

so that

$$\mathbf{X} = c_1 \begin{pmatrix} 0 \\ 1 \\ 1 \end{pmatrix} e^t + c_2 \left[\begin{pmatrix} 0 \\ 1 \\ 1 \end{pmatrix} te^t + \begin{pmatrix} 0 \\ 1 \\ 0 \end{pmatrix} e^t \right] + c_3 \left[\begin{pmatrix} 0 \\ 1 \\ 1 \end{pmatrix} \frac{t^2}{2}e^t + \begin{pmatrix} 0 \\ 1 \\ 0 \end{pmatrix} te^t + \begin{pmatrix} 1/2 \\ 0 \\ 0 \end{pmatrix} e^t \right].$$

30. We have $\det(\mathbf{A} - \lambda\mathbf{I}) = -(\lambda + 1)(\lambda - 1)^2 = 0$. For $\lambda_1 = -1$ we obtain

$$\mathbf{K}_1 = \begin{pmatrix} -1 \\ 0 \\ 1 \end{pmatrix}.$$

For $\lambda_2 = 1$ we obtain

$$\mathbf{K}_2 = \begin{pmatrix} 1 \\ 0 \\ 1 \end{pmatrix} \quad \text{and} \quad \mathbf{K}_3 = \begin{pmatrix} 0 \\ 1 \\ 0 \end{pmatrix}$$

so that

$$\mathbf{X} = c_1 \begin{pmatrix} -1 \\ 0 \\ 1 \end{pmatrix} e^{-t} + c_2 \begin{pmatrix} 1 \\ 0 \\ 1 \end{pmatrix} e^{t} + c_3 \begin{pmatrix} 0 \\ 1 \\ 0 \end{pmatrix} e^{t}.$$

If

$$\mathbf{X}(0) = \begin{pmatrix} 1 \\ 2 \\ 5 \end{pmatrix}$$

then $c_1 = 2$, $c_2 = 3$, and $c_3 = 2$.

In Problems 33-45 the form of the answer will vary according to the choice of eigenvector. For example, in Problem 33, if $\mathbf{K}_1$ is chosen to be $\begin{pmatrix} 1 \\ 2 - i \end{pmatrix}$ the solution has the form

$$\mathbf{X} = c_1 \begin{pmatrix} \cos t \\ 2\cos t + \sin t \end{pmatrix} e^{4t} + c_2 \begin{pmatrix} \sin t \\ 2\sin t - \cos t \end{pmatrix} e^{4t}.$$

The surest way to check your answer, if it disagrees with the one in the answer section of the text or this manual, is to substitute your answer into the system of DEs.

33. We have $\det(\mathbf{A} - \lambda \mathbf{I}) = \lambda^2 - 8\lambda + 17 = 0$. For $\lambda_1 = 4 + i$ we obtain

$$\mathbf{K}_1 = \begin{pmatrix} 2 + i \\ 5 \end{pmatrix}$$

so that

$$\mathbf{X}_1 = \begin{pmatrix} 2 + i \\ 5 \end{pmatrix} e^{(4+i)t} = \begin{pmatrix} 2\cos t - \sin t \\ 5\cos t \end{pmatrix} e^{4t} + i \begin{pmatrix} \cos t + 2\sin t \\ 5\sin t \end{pmatrix} e^{4t}.$$

Then

$$\mathbf{X} = c_1 \begin{pmatrix} 2\cos t - \sin t \\ 5\cos t \end{pmatrix} e^{4t} + c_2 \begin{pmatrix} \cos t + 2\sin t \\ 5\sin t \end{pmatrix} e^{4t}.$$

36. We have $\det(\mathbf{A} - \lambda \mathbf{I}) = \lambda^2 - 10\lambda + 34 = 0$. For $\lambda_1 = 5 + 3i$ we obtain

$$\mathbf{K}_1 = \begin{pmatrix} 1 - 3i \\ 2 \end{pmatrix}$$

so that

$$\mathbf{X}_1 = \begin{pmatrix} 1 - 3i \\ 2 \end{pmatrix} e^{(5+3i)t} = \begin{pmatrix} \cos 3t + 3\sin 3t \\ 2\cos 3t \end{pmatrix} e^{5t} + i \begin{pmatrix} \sin 3t - 3\cos 3t \\ 2\sin 3t \end{pmatrix} e^{5t}.$$

Then

$$\mathbf{X} = c_1 \begin{pmatrix} \cos 3t + 3\sin 3t \\ 2\cos 3t \end{pmatrix} e^{5t} + c_2 \begin{pmatrix} \sin 3t - 3\cos 3t \\ 2\sin 3t \end{pmatrix} e^{5t}.$$

39. We have $\det(\mathbf{A} - \lambda\mathbf{I}) = -\lambda\left(\lambda^2 + 1\right) = 0$. For $\lambda_1 = 0$ we obtain

$$\mathbf{K}_1 = \begin{pmatrix} 1 \\ 0 \\ 0 \end{pmatrix}.$$

For $\lambda_2 = i$ we obtain

$$\mathbf{K}_2 = \begin{pmatrix} -i \\ i \\ 1 \end{pmatrix}$$

so that

$$\mathbf{X}_2 = \begin{pmatrix} -i \\ i \\ 1 \end{pmatrix} e^{it} = \begin{pmatrix} \sin t \\ -\sin t \\ \cos t \end{pmatrix} + i \begin{pmatrix} -\cos t \\ \cos t \\ \sin t \end{pmatrix}.$$

Then

$$\mathbf{X} = c_1 \begin{pmatrix} 1 \\ 0 \\ 0 \end{pmatrix} + c_2 \begin{pmatrix} \sin t \\ -\sin t \\ \cos t \end{pmatrix} + c_3 \begin{pmatrix} -\cos t \\ \cos t \\ \sin t \end{pmatrix}.$$

42. We have $\det(\mathbf{A} - \lambda\mathbf{I}) = -(\lambda - 6)(\lambda^2 - 8\lambda + 20) = 0$. For $\lambda_1 = 6$ we obtain

$$\mathbf{K}_1 = \begin{pmatrix} 0 \\ 1 \\ 0 \end{pmatrix}.$$

For $\lambda_2 = 4 + 2i$ we obtain

$$\mathbf{K}_2 = \begin{pmatrix} -i \\ 0 \\ 2 \end{pmatrix}.$$

so that

$$\mathbf{X}_2 = \begin{pmatrix} -i \\ 0 \\ 2 \end{pmatrix} e^{(4+2i)t} = \begin{pmatrix} \sin 2t \\ 0 \\ 2\cos 2t \end{pmatrix} e^{4t} + i \begin{pmatrix} -\cos 2t \\ 0 \\ 2\sin 2t \end{pmatrix} e^{4t}.$$

Then

$$\mathbf{X} = c_1 \begin{pmatrix} 0 \\ 1 \\ 0 \end{pmatrix} e^{6t} + c_2 \begin{pmatrix} \sin 2t \\ 0 \\ 2\cos 2t \end{pmatrix} e^{4t} + c_3 \begin{pmatrix} -\cos 2t \\ 0 \\ 2\sin 2t \end{pmatrix} e^{4t}.$$

45. We have $\det(\mathbf{A} - \lambda\mathbf{I}) = (1 - \lambda)(\lambda^2 + 25) = 0$. For $\lambda_1 = 1$ we obtain

$$\mathbf{K}_1 = \begin{pmatrix} 25 \\ -7 \\ 6 \end{pmatrix}.$$

For $\lambda_2 = 5i$ we obtain

$$\mathbf{K}_2 = \begin{pmatrix} 1 + 5i \\ 1 \\ 1 \end{pmatrix}$$

so that

$$\mathbf{X}_2 = \begin{pmatrix} 1 + 5i \\ 1 \\ 1 \end{pmatrix} e^{5it} = \begin{pmatrix} \cos 5t - 5\sin 5t \\ \cos 5t \\ \cos 5t \end{pmatrix} + i \begin{pmatrix} \sin 5t + 5\cos 5t \\ \sin 5t \\ \sin 5t \end{pmatrix}.$$

Then

$$\mathbf{X} = c_1 \begin{pmatrix} 25 \\ -7 \\ 6 \end{pmatrix} e^t + c_2 \begin{pmatrix} \cos 5t - 5\sin 5t \\ \cos 5t \\ \cos 5t \end{pmatrix} + c_3 \begin{pmatrix} \sin 5t + 5\cos 5t \\ \sin 5t \\ \sin 5t \end{pmatrix}.$$

If

$$\mathbf{X}(0) = \begin{pmatrix} 4 \\ 6 \\ -7 \end{pmatrix}$$

then $c_1 = c_2 = -1$ and $c_3 = 6$.

Section 8.3   Nonhomogeneous Linear Systems

The terminology and concepts listed below provide an outline of the main ideas encountered in this section. These can be useful when preparing for a quiz or test.

Terminology and Concepts

- method of undetermined coefficients for finding a particular solution of a nonhomogeneous linear system of DEs

- fundamental matrix of a linear system of DEs

- method of variation of parameters for finding a particular solution of a nonhomogeneous linear system of DEs

The basic skills listed below summarize the more mechanical types of problems encountered in the exercise set for this section.

Basic Skills

- use undetermined coefficients to solve a nonhomogeneous linear system of DEs
- use variation of parameters to solve a nonhomogeneous linear system of DEs

Exercises 8.3 *Hints, Suggestions, Solutions, and Examples*

3. Solving

$$\det(\mathbf{A} - \lambda\mathbf{I}) = \begin{vmatrix} 1 - \lambda & 3 \\ 3 & 1 - \lambda \end{vmatrix} = \lambda^2 - 2\lambda - 8 = (\lambda - 4)(\lambda + 2) = 0$$

we obtain eigenvalues $\lambda_1 = -2$ and $\lambda_2 = 4$. Corresponding eigenvectors are

$$\mathbf{K}_1 = \begin{pmatrix} 1 \\ -1 \end{pmatrix} \quad \text{and} \quad \mathbf{K}_2 = \begin{pmatrix} 1 \\ 1 \end{pmatrix}.$$

Thus

$$\mathbf{X}_c = c_1 \begin{pmatrix} 1 \\ -1 \end{pmatrix} e^{-2t} + c_2 \begin{pmatrix} 1 \\ 1 \end{pmatrix} e^{4t}.$$

Substituting

$$\mathbf{X}_p = \begin{pmatrix} a_3 \\ b_3 \end{pmatrix} t^2 + \begin{pmatrix} a_2 \\ b_2 \end{pmatrix} t + \begin{pmatrix} a_1 \\ b_1 \end{pmatrix}$$

into the system yields

$$a_3 + 3b_3 = 2 \qquad a_2 + 3b_2 = 2a_3 \qquad a_1 + 3b_1 = a_2$$

$$3a_3 + b_3 = 0 \qquad 3a_2 + b_2 + 1 = 2b_3 \qquad 3a_1 + b_1 + 5 = b_2$$

from which we obtain $a_3 = -1/4$, $b_3 = 3/4$, $a_2 = 1/4$, $b_2 = -1/4$, $a_1 = -2$, and $b_1 = 3/4$. Then

$$\mathbf{X}(t) = c_1 \begin{pmatrix} 1 \\ -1 \end{pmatrix} e^{-2t} + c_2 \begin{pmatrix} 1 \\ 1 \end{pmatrix} e^{4t} + \begin{pmatrix} -1/4 \\ 3/4 \end{pmatrix} t^2 + \begin{pmatrix} 1/4 \\ -1/4 \end{pmatrix} t + \begin{pmatrix} -2 \\ 3/4 \end{pmatrix}.$$

6. Solving

$$\det(\mathbf{A} - \lambda\mathbf{I}) = \begin{vmatrix} -1 - \lambda & 5 \\ -1 & 1 - \lambda \end{vmatrix} = \lambda^2 + 4 = 0$$

we obtain the eigenvalues $\lambda_1 = 2i$ and $\lambda_2 = -2i$. Corresponding eigenvectors are

$$\mathbf{K}_1 = \begin{pmatrix} 5 \\ 1 + 2i \end{pmatrix} \quad \text{and} \quad \mathbf{K}_2 = \begin{pmatrix} 5 \\ 1 - 2i \end{pmatrix}.$$

Thus

$$\mathbf{X}_c = c_1 \begin{pmatrix} 5\cos 2t \\ \cos 2t - 2\sin 2t \end{pmatrix} + c_2 \begin{pmatrix} 5\sin 2t \\ 2\cos 2t + \sin 2t \end{pmatrix}.$$

Substituting

$$\mathbf{X}_p = \begin{pmatrix} a_2 \\ b_2 \end{pmatrix} \cos t + \begin{pmatrix} a_1 \\ b_1 \end{pmatrix} \sin t$$

into the system yields

$$-a_2 + 5b_2 - a_1 = 0$$

$$-a_2 + b_2 - b_1 - 2 = 0$$

$$-a_1 + 5b_1 + a_2 + 1 = 0$$

$$-a_1 + b_1 + b_2 = 0$$

from which we obtain $a_2 = -3$, $b_2 = -2/3$, $a_1 = -1/3$, and $b_1 = 1/3$. Then

$$\mathbf{X}(t) = c_1 \begin{pmatrix} 5\cos 2t \\ \cos 2t - 2\sin 2t \end{pmatrix} + c_2 \begin{pmatrix} 5\sin 2t \\ 2\cos 2t + \sin 2t \end{pmatrix} + \begin{pmatrix} -3 \\ -2/3 \end{pmatrix} \cos t + \begin{pmatrix} -1/3 \\ 1/3 \end{pmatrix} \sin t.$$

9. Solving

$$\det(\mathbf{A} - \lambda\mathbf{I}) = \begin{vmatrix} -1 - \lambda & -2 \\ 3 & 4 - \lambda \end{vmatrix} = \lambda^2 - 3\lambda + 2 = (\lambda - 1)(\lambda - 2) = 0$$

we obtain the eigenvalues $\lambda_1 = 1$ and $\lambda_2 = 2$. Corresponding eigenvectors are

$$\mathbf{K}_1 = \begin{pmatrix} 1 \\ -1 \end{pmatrix}, \quad \text{and} \quad \mathbf{K}_2 = \begin{pmatrix} -4 \\ 6 \end{pmatrix}.$$

Thus

$$\mathbf{X}_c = c_1 \begin{pmatrix} 1 \\ -1 \end{pmatrix} e^t + c_2 \begin{pmatrix} -4 \\ 6 \end{pmatrix} e^{2t}.$$

Substituting

$$\mathbf{X}_p = \begin{pmatrix} a_1 \\ b_1 \end{pmatrix}$$

into the system yields

$$-a_1 - 2b_1 = -3$$

$$3a_1 + 4b_1 = -3$$

from which we obtain $a_1 = -9$ and $b_1 = 6$. Then

$$\mathbf{X}(t) = c_1 \begin{pmatrix} 1 \\ -1 \end{pmatrix} e^t + c_2 \begin{pmatrix} -4 \\ 6 \end{pmatrix} e^{2t} + \begin{pmatrix} -9 \\ 6 \end{pmatrix}.$$

Setting

$$\mathbf{X}(0) = \begin{pmatrix} -4 \\ 5 \end{pmatrix}$$

we obtain

$$c_1 - 4c_2 - 9 = -4$$

$$-c_1 + 6c_2 + 6 = 5.$$

Then $c_1 = 13$ and $c_2 = 2$ so

$$\mathbf{X}(t) = 13 \begin{pmatrix} 1 \\ -1 \end{pmatrix} e^t + 2 \begin{pmatrix} -4 \\ 6 \end{pmatrix} e^{2t} + \begin{pmatrix} -9 \\ 6 \end{pmatrix}.$$

12. From

$$\mathbf{X}' = \begin{pmatrix} 2 & -1 \\ 3 & -2 \end{pmatrix} \mathbf{X} + \begin{pmatrix} 0 \\ 4 \end{pmatrix} t$$

we obtain

$$\mathbf{X}_c = c_1 \begin{pmatrix} 1 \\ 1 \end{pmatrix} e^t + c_2 \begin{pmatrix} 1 \\ 3 \end{pmatrix} e^{-t}.$$

Then

$$\mathbf{\Phi} = \begin{pmatrix} e^t & e^{-t} \\ e^t & 3e^{-t} \end{pmatrix} \quad \text{and} \quad \mathbf{\Phi}^{-1} = \begin{pmatrix} \frac{3}{2}e^{-t} & -\frac{1}{2}e^{-t} \\ -\frac{1}{2}e^t & \frac{1}{2}e^t \end{pmatrix}$$

so that

$$\mathbf{U} = \int \mathbf{\Phi}^{-1}\mathbf{F}\,dt = \int \begin{pmatrix} -2te^{-t} \\ 2te^t \end{pmatrix} dt = \begin{pmatrix} 2te^{-t} + 2e^{-t} \\ 2te^t - 2e^t \end{pmatrix}$$

and

$$\mathbf{X}_p = \mathbf{\Phi}\mathbf{U} = \begin{pmatrix} 4 \\ 8 \end{pmatrix} t + \begin{pmatrix} 0 \\ -4 \end{pmatrix}.$$

15. From

$$\mathbf{X}' = \begin{pmatrix} 0 & 2 \\ -1 & 3 \end{pmatrix} \mathbf{X} + \begin{pmatrix} 1 \\ -1 \end{pmatrix} e^t$$

we obtain

$$\mathbf{X}_c = c_1 \begin{pmatrix} 2 \\ 1 \end{pmatrix} e^t + c_2 \begin{pmatrix} 1 \\ 1 \end{pmatrix} e^{2t}.$$

Then

$$\mathbf{\Phi} = \begin{pmatrix} 2e^t & e^{2t} \\ e^t & e^{2t} \end{pmatrix} \quad \text{and} \quad \mathbf{\Phi}^{-1} = \begin{pmatrix} e^{-t} & -e^{-t} \\ -e^{-2t} & 2e^{-2t} \end{pmatrix}$$

so that

$$\mathbf{U} = \int \mathbf{\Phi}^{-1}\mathbf{F}\,dt = \int \begin{pmatrix} 2 \\ -3e^{-t} \end{pmatrix} dt = \begin{pmatrix} 2t \\ 3e^{-t} \end{pmatrix}$$

and

$$\mathbf{X}_p = \mathbf{\Phi}\mathbf{U} = \begin{pmatrix} 4 \\ 2 \end{pmatrix} te^t + \begin{pmatrix} 3 \\ 3 \end{pmatrix} e^t.$$

18. From

$$\mathbf{X}' = \begin{pmatrix} 1 & 8 \\ 1 & -1 \end{pmatrix} \mathbf{X} + \begin{pmatrix} e^{-t} \\ te^t \end{pmatrix}$$

we obtain

$$\mathbf{X}_c = c_1 \begin{pmatrix} 4 \\ 1 \end{pmatrix} e^{3t} + c_2 \begin{pmatrix} -2 \\ 1 \end{pmatrix} e^{-3t}.$$

Then

$$\mathbf{\Phi} = \begin{pmatrix} 4e^{3t} & -2e^{3t} \\ e^{3t} & e^{-3t} \end{pmatrix} \quad \text{and} \quad \mathbf{\Phi}^{-1} = \begin{pmatrix} \frac{1}{6}e^{-3t} & \frac{1}{3}e^{-3t} \\ -\frac{1}{6}e^{3t} & \frac{2}{3}e^{3t} \end{pmatrix}$$

so that

$$\mathbf{U} = \int \mathbf{\Phi}^{-1}\mathbf{F}\,dt = \int \begin{pmatrix} \frac{1}{6}e^{-4t} + \frac{1}{3}te^{-2t} \\ -\frac{1}{6}e^{2t} + \frac{2}{3}te^{4t} \end{pmatrix} dt = \begin{pmatrix} -\frac{1}{24}e^{-4t} - \frac{1}{6}te^{-2t} - \frac{1}{12}e^{-2t} \\ -\frac{1}{12}e^{2t} + \frac{1}{6}te^{4t} - \frac{1}{24}e^{4t} \end{pmatrix}$$

and

$$\mathbf{X}_p = \mathbf{\Phi U} = \begin{pmatrix} -te^{t} - \frac{1}{4}e^{t} \\ -\frac{1}{8}e^{-t} - \frac{1}{8}e^{t} \end{pmatrix}.$$

21. From

$$\mathbf{X}' = \begin{pmatrix} 0 & -1 \\ 1 & 0 \end{pmatrix} \mathbf{X} + \begin{pmatrix} \sec t \\ 0 \end{pmatrix}$$

we obtain

$$\mathbf{X}_c = c_1 \begin{pmatrix} \cos t \\ \sin t \end{pmatrix} + c_2 \begin{pmatrix} \sin t \\ -\cos t \end{pmatrix}.$$

Then

$$\mathbf{\Phi} = \begin{pmatrix} \cos t & \sin t \\ \sin t & -\cos t \end{pmatrix} \quad \text{and} \quad \mathbf{\Phi}^{-1} = \begin{pmatrix} \cos t & \sin t \\ \sin t & -\cos t \end{pmatrix}$$

so that

$$\mathbf{U} = \int \mathbf{\Phi}^{-1}\mathbf{F}\,dt = \int \begin{pmatrix} 1 \\ \tan t \end{pmatrix} dt = \begin{pmatrix} t \\ -\ln|\cos t| \end{pmatrix}$$

and

$$\mathbf{X}_p = \mathbf{\Phi U} = \begin{pmatrix} t\cos t - \sin t \ln|\cos t| \\ t\sin t + \cos t \ln|\cos t| \end{pmatrix}.$$

24. From

$$\mathbf{X}' = \begin{pmatrix} 2 & -2 \\ 8 & -6 \end{pmatrix} \mathbf{X} + \begin{pmatrix} 1 \\ 3 \end{pmatrix} \frac{1}{t} e^{-2t}$$

we obtain

$$\mathbf{X}_c = c_1 \begin{pmatrix} 1 \\ 2 \end{pmatrix} e^{-2t} + c_2 \left[\begin{pmatrix} 1 \\ 2 \end{pmatrix} te^{-2t} + \begin{pmatrix} 1/2 \\ 1/2 \end{pmatrix} e^{-2t} \right].$$

Then

$$\mathbf{\Phi} = \begin{pmatrix} 1 & t+\frac{1}{2} \\ 2 & 2t+\frac{1}{2} \end{pmatrix} e^{-2t} \quad \text{and} \quad \mathbf{\Phi}^{-1} = \begin{pmatrix} -4t-1 & 2t+1 \\ 4 & -2 \end{pmatrix} e^{2t}$$

so that

$$\mathbf{U} = \int \mathbf{\Phi}^{-1}\mathbf{F}\,dt = \int \begin{pmatrix} 2+2/t \\ -2/t \end{pmatrix} dt = \begin{pmatrix} 2t + 2\ln t \\ -2\ln t \end{pmatrix}$$

and

$$\mathbf{X}_p = \mathbf{\Phi U} = \begin{pmatrix} 2t + \ln t - 2t\ln t \\ 4t + 3\ln t - 4t\ln t \end{pmatrix} e^{-2t}.$$

27. From

$$\mathbf{X'} = \begin{pmatrix} 1 & 2 \\ -1/2 & 1 \end{pmatrix} \mathbf{X} + \begin{pmatrix} \csc t \\ \sec t \end{pmatrix} e^t$$

we obtain

$$\mathbf{X}_c = c_1 \begin{pmatrix} 2\sin t \\ \cos t \end{pmatrix} e^t + c_2 \begin{pmatrix} 2\cos t \\ -\sin t \end{pmatrix} e^t.$$

Then

$$\mathbf{\Phi} = \begin{pmatrix} 2\sin t & 2\cos t \\ \cos t & -\sin t \end{pmatrix} e^t \quad \text{and} \quad \mathbf{\Phi}^{-1} = \begin{pmatrix} \frac{1}{2}\sin t & \cos t \\ \frac{1}{2}\cos t & -\sin t \end{pmatrix} e^{-t}.$$

so that

$$\mathbf{U} = \int \mathbf{\Phi}^{-1}\mathbf{F}\,dt = \int \begin{pmatrix} \frac{3}{2} \\ \frac{1}{2}\cot t - \tan t \end{pmatrix} dt = \begin{pmatrix} \frac{3}{2}t \\ \frac{1}{2}\ln|\sin t| + \ln|\cos t| \end{pmatrix}$$

and

$$\mathbf{X}_p = \mathbf{\Phi U} = \begin{pmatrix} 3\sin t \\ \frac{3}{2}\cos t \end{pmatrix} te^t + \begin{pmatrix} \cos t \\ -\frac{1}{2}\sin t \end{pmatrix} e^t \ln|\sin t| + \begin{pmatrix} 2\cos t \\ -\sin t \end{pmatrix} e^t \ln|\cos t|.$$

30. From

$$\mathbf{X'} = \begin{pmatrix} 3 & -1 & -1 \\ 1 & 1 & -1 \\ 1 & -1 & 1 \end{pmatrix} \mathbf{X} + \begin{pmatrix} 0 \\ t \\ 2e^t \end{pmatrix}$$

we obtain

$$\mathbf{X}_c = c_1 \begin{pmatrix} 1 \\ 1 \\ 1 \end{pmatrix} e^t + c_2 \begin{pmatrix} 1 \\ 1 \\ 0 \end{pmatrix} e^{2t} + c_3 \begin{pmatrix} 1 \\ 0 \\ 1 \end{pmatrix} e^{2t}.$$

Then

$$\mathbf{\Phi} = \begin{pmatrix} e^t & e^{2t} & e^{2t} \\ e^t & e^{2t} & 0 \\ e^t & 0 & e^{2t} \end{pmatrix} \quad \text{and} \quad \mathbf{\Phi}^{-1} = \begin{pmatrix} -e^{-t} & e^{-t} & e^{-t} \\ e^{-2t} & 0 & -e^{-2t} \\ e^{-2t} & -e^{-2t} & 0 \end{pmatrix}$$

so that

$$\mathbf{U} = \int \mathbf{\Phi}^{-1}\mathbf{F}\,dt = \int \begin{pmatrix} te^{-t} + 2 \\ -2e^{-t} \\ -te^{-2t} \end{pmatrix} dt = \begin{pmatrix} -te^{-t} - e^{-t} + 2t \\ 2e^{-t} \\ \frac{1}{2}te^{-2t} + \frac{1}{4}e^{-2t} \end{pmatrix}$$

and

$$\mathbf{X}_p = \mathbf{\Phi U} = \begin{pmatrix} -1/2 \\ -1 \\ -1/2 \end{pmatrix} t + \begin{pmatrix} -3/4 \\ -1 \\ -3/4 \end{pmatrix} + \begin{pmatrix} 2 \\ 2 \\ 0 \end{pmatrix} e^t + \begin{pmatrix} 2 \\ 2 \\ 2 \end{pmatrix} te^t.$$

33. Let $\mathbf{I} = \begin{pmatrix} i_1 \\ i_2 \end{pmatrix}$ so that

$$\mathbf{I}' = \begin{pmatrix} -11 & 3 \\ 3 & -3 \end{pmatrix} \mathbf{I} + \begin{pmatrix} 100\sin t \\ 0 \end{pmatrix}$$

and

$$\mathbf{I}_c = c_1 \begin{pmatrix} 1 \\ 3 \end{pmatrix} e^{-2t} + c_2 \begin{pmatrix} 3 \\ -1 \end{pmatrix} e^{-12t}.$$

Then

$$\boldsymbol{\Phi} = \begin{pmatrix} e^{-2t} & 3e^{-12t} \\ 3e^{-2t} & -e^{-12t} \end{pmatrix}, \quad \boldsymbol{\Phi}^{-1} = \begin{pmatrix} \frac{1}{10}e^{2t} & \frac{3}{10}e^{2t} \\ \frac{3}{10}e^{12t} & -\frac{1}{10}e^{12t} \end{pmatrix},$$

$$\mathbf{U} = \int \boldsymbol{\Phi}^{-1}\mathbf{F}\,dt = \int \begin{pmatrix} 10e^{2t}\sin t \\ 30e^{12t}\sin t \end{pmatrix} dt = \begin{pmatrix} 2e^{2t}(2\sin t - \cos t) \\ \frac{6}{29}e^{12t}(12\sin t - \cos t) \end{pmatrix},$$

and

$$\mathbf{I}_p = \boldsymbol{\Phi}\mathbf{U} = \begin{pmatrix} \frac{332}{29}\sin t - \frac{76}{29}\cos t \\ \frac{276}{29}\sin t - \frac{168}{29}\cos t \end{pmatrix}$$

so that

$$\mathbf{I} = c_1 \begin{pmatrix} 1 \\ 3 \end{pmatrix} e^{-2t} + c_2 \begin{pmatrix} 3 \\ -1 \end{pmatrix} e^{-12t} + \mathbf{I}_p.$$

If $\mathbf{I}(0) = \begin{pmatrix} 0 \\ 0 \end{pmatrix}$ then $c_1 = 2$ and $c_2 = \frac{6}{29}$.

Section 8.4   Matrix Exponential

The terminology and concepts listed below provide an outline of the main ideas encountered in this section. These can be useful when preparing for a quiz or test.

Terminology and Concepts

- infinite series representation of the matrix exponential $e^{\mathbf{A}t}$
- the matrix exponential is a fundamental matrix

The basic skills listed below summarize the more mechanical types of problems encountered in the exercise set for this section.

Basic Skills

- compute $e^{\mathbf{A}t}$ for a given matrix $\mathbf{A}$
- use the matrix exponential to solve a nonhomogeneous linear system

Use of Computers To compute the matrix exponential for a square matrix $\mathbf{A}t$ use

MatrixExp[At]   (*Mathematica*)

with(linalg):   (*Maple*)
exponential(A,t);

expm(At)   (MATLAB)

Exercises 8.4 *Hints, Suggestions, Solutions, and Examples*

3. For

$$\mathbf{A} = \begin{pmatrix} 1 & 1 & 1 \\ 1 & 1 & 1 \\ -2 & -2 & -2 \end{pmatrix}$$

we have

$$\mathbf{A}^2 = \begin{pmatrix} 1 & 1 & 1 \\ 1 & 1 & 1 \\ -2 & -2 & -2 \end{pmatrix} \begin{pmatrix} 1 & 1 & 1 \\ 1 & 1 & 1 \\ -2 & -2 & -2 \end{pmatrix} = \begin{pmatrix} 0 & 0 & 0 \\ 0 & 0 & 0 \\ 0 & 0 & 0 \end{pmatrix}.$$

Thus, $\mathbf{A}^3 = \mathbf{A}^4 = \mathbf{A}^5 = \cdots = \mathbf{0}$ and

$$e^{\mathbf{A}t} = \mathbf{I} + \mathbf{A}t = \begin{pmatrix} 1 & 0 & 0 \\ 0 & 1 & 0 \\ 0 & 0 & 1 \end{pmatrix} + \begin{pmatrix} t & t & t \\ t & t & t \\ -2t & -2t & -2t \end{pmatrix} = \begin{pmatrix} t+1 & t & t \\ t & t+1 & t \\ -2t & -2t & -2t+1 \end{pmatrix}.$$

6. In Problem 2 it is shown that for

$$\mathbf{A} = \begin{pmatrix} 0 & 1 \\ 1 & 0 \end{pmatrix}$$

we have

$$e^{\mathbf{A}t} = \begin{pmatrix} \cosh t & \sinh t \\ \sinh t & \cosh t \end{pmatrix}.$$

Thus, the solution of the given system is

$$\mathbf{X} = \begin{pmatrix} \cosh t & \sinh t \\ \sinh t & \cosh t \end{pmatrix} \begin{pmatrix} c_1 \\ c_2 \end{pmatrix} = c_1 \begin{pmatrix} \cosh t \\ \sinh t \end{pmatrix} + c_2 \begin{pmatrix} \sinh t \\ \cosh t \end{pmatrix}.$$

9. To solve

$$\mathbf{X}' = \begin{pmatrix} 1 & 0 \\ 0 & 2 \end{pmatrix} \mathbf{X} + \begin{pmatrix} 3 \\ -1 \end{pmatrix}$$

we identify $t_0 = 0$, $\mathbf{F}(t) = \begin{pmatrix} 3 \\ -1 \end{pmatrix}$, and use the results of Problem 1 and equation (5) in the text.

$$\mathbf{X}(t) = e^{\mathbf{A}t}\mathbf{C} + e^{\mathbf{A}t} \int_{t_0}^{t} e^{-\mathbf{A}s}\mathbf{F}(s)\,ds$$

$$= \begin{pmatrix} e^t & 0 \\ 0 & e^{2t} \end{pmatrix} \begin{pmatrix} c_1 \\ c_2 \end{pmatrix} + \begin{pmatrix} e^t & 0 \\ 0 & e^{2t} \end{pmatrix} \int_0^t \begin{pmatrix} e^{-s} & 0 \\ 0 & e^{-2s} \end{pmatrix} \begin{pmatrix} 3 \\ -1 \end{pmatrix} ds$$

$$= \begin{pmatrix} c_1 e^t \\ c_2 e^{2t} \end{pmatrix} + \begin{pmatrix} e^t & 0 \\ 0 & e^{2t} \end{pmatrix} \int_0^t \begin{pmatrix} 3e^{-s} \\ -e^{-2s} \end{pmatrix} ds$$

$$= \begin{pmatrix} c_1 e^t \\ c_2 e^{2t} \end{pmatrix} + \begin{pmatrix} e^t & 0 \\ 0 & e^{2t} \end{pmatrix} \begin{pmatrix} -3e^{-s} \\ \frac{1}{2}e^{-2s} \end{pmatrix} \Big|_0^t$$

$$= \begin{pmatrix} c_1 e^t \\ c_2 e^{2t} \end{pmatrix} + \begin{pmatrix} e^t & 0 \\ 0 & e^{2t} \end{pmatrix} \begin{pmatrix} -3e^{-t} + 3 \\ \frac{1}{2}e^{-2t} - \frac{1}{2} \end{pmatrix}$$

$$= \begin{pmatrix} c_1 e^t \\ c_2 e^{2t} \end{pmatrix} + \begin{pmatrix} -3 + 3e^t \\ \frac{1}{2} - \frac{1}{2}e^{2t} \end{pmatrix} = c_3 \begin{pmatrix} 1 \\ 0 \end{pmatrix} e^t + c_4 \begin{pmatrix} 0 \\ 1 \end{pmatrix} e^{2t} + \begin{pmatrix} -3 \\ \frac{1}{2} \end{pmatrix}.$$

12. We use the result of Problem 6 above. To solve

$$\mathbf{X}' = \begin{pmatrix} 0 & 1 \\ 1 & 0 \end{pmatrix} \mathbf{X} + \begin{pmatrix} \cosh t \\ \sinh t \end{pmatrix}$$

we identify $t_0 = 0$, $\mathbf{F}(t) = \begin{pmatrix} \cosh t \\ \sinh t \end{pmatrix}$, and use the results of Problem 2 and equation (5) in the text.

$$\mathbf{X}(t) = e^{\mathbf{A}t}\mathbf{C} + e^{\mathbf{A}t} \int_{t_0}^{t} e^{-\mathbf{A}s}\mathbf{F}(s)\,ds$$

$$= \begin{pmatrix} \cosh t & \sinh t \\ \sinh t & \cosh t \end{pmatrix} \begin{pmatrix} c_1 \\ c_2 \end{pmatrix} + \begin{pmatrix} \cosh t & \sinh t \\ \sinh t & \cosh t \end{pmatrix} \int_0^t \begin{pmatrix} \cosh s & -\sinh s \\ -\sinh s & \cosh s \end{pmatrix} \begin{pmatrix} \cosh s \\ \sinh s \end{pmatrix} ds$$

$$= \begin{pmatrix} c_1 \cosh t + c_2 \sinh t \\ c_1 \sinh t + c_2 \cosh t \end{pmatrix} + \begin{pmatrix} \cosh t & \sinh t \\ \sinh t & \cosh t \end{pmatrix} \int_0^t \begin{pmatrix} 1 \\ 0 \end{pmatrix} ds$$

$$= \begin{pmatrix} c_1 \cosh t + c_2 \sinh t \\ c_1 \sinh t + c_2 \cosh t \end{pmatrix} + \begin{pmatrix} \cosh t & \sinh t \\ \sinh t & \cosh t \end{pmatrix} \begin{pmatrix} s \\ 0 \end{pmatrix} \Big|_0^t$$

$$= \begin{pmatrix} c_1 \cosh t + c_2 \sinh t \\ c_1 \sinh t + c_2 \cosh t \end{pmatrix} + \begin{pmatrix} \cosh t & \sinh t \\ \sinh t & \cosh t \end{pmatrix} \begin{pmatrix} t \\ 0 \end{pmatrix}$$

$$= \begin{pmatrix} c_1 \cosh t + c_2 \sinh t \\ c_1 \sinh t + c_2 \cosh t \end{pmatrix} + \begin{pmatrix} t \cosh t \\ t \sinh t \end{pmatrix} = c_1 \begin{pmatrix} \cosh t \\ \sinh t \end{pmatrix} + c_2 \begin{pmatrix} \sinh t \\ \cosh t \end{pmatrix} + t \begin{pmatrix} \cosh t \\ \sinh t \end{pmatrix}.$$

15. From $s\mathbf{I} - \mathbf{A} = \begin{pmatrix} s-4 & -3 \\ 4 & s+4 \end{pmatrix}$ we find

$$(s\mathbf{I} - \mathbf{A})^{-1} = \begin{pmatrix} \dfrac{3/2}{s-2} - \dfrac{1/2}{s+2} & \dfrac{3/4}{s-2} - \dfrac{3/4}{s+2} \\[2ex] \dfrac{-1}{s-2} + \dfrac{1}{s+2} & \dfrac{-1/2}{s-2} + \dfrac{3/2}{s+2} \end{pmatrix}$$

and

$$e^{\mathbf{A}t} = \begin{pmatrix} \frac{3}{2}e^{2t} - \frac{1}{2}e^{-2t} & \frac{3}{4}e^{2t} - \frac{3}{4}e^{-2t} \\[1ex] -e^{2t} + e^{-2t} & -\frac{1}{2}e^{2t} + \frac{3}{2}e^{-2t} \end{pmatrix}.$$

The general solution of the system is then

$$\mathbf{X} = e^{\mathbf{A}t}\mathbf{C} = \begin{pmatrix} \frac{3}{2}e^{2t} - \frac{1}{2}e^{-2t} & \frac{3}{4}e^{2t} - \frac{3}{4}e^{-2t} \\[1ex] -e^{2t} + e^{-2t} & -\frac{1}{2}e^{2t} + \frac{3}{2}e^{-2t} \end{pmatrix} \begin{pmatrix} c_1 \\ c_2 \end{pmatrix}$$

$$= c_1 \begin{pmatrix} 3/2 \\ -1 \end{pmatrix} e^{2t} + c_1 \begin{pmatrix} -1/2 \\ 1 \end{pmatrix} e^{-2t} + c_2 \begin{pmatrix} 3/4 \\ -1/2 \end{pmatrix} e^{2t} + c_2 \begin{pmatrix} -3/4 \\ 3/2 \end{pmatrix} e^{-2t}$$

$$= \left(\frac{1}{2}c_1 + \frac{1}{4}c_2\right) \begin{pmatrix} 3 \\ -2 \end{pmatrix} e^{2t} + \left(-\frac{1}{2}c_1 - \frac{3}{4}c_2\right) \begin{pmatrix} 1 \\ -2 \end{pmatrix} e^{-2t}$$

$$= c_3 \begin{pmatrix} 3 \\ -2 \end{pmatrix} e^{2t} + c_4 \begin{pmatrix} 1 \\ -2 \end{pmatrix} e^{-2t}.$$

18. From $s\mathbf{I} - \mathbf{A} = \begin{pmatrix} s & -1 \\ 2 & s+2 \end{pmatrix}$ we find

$$(s\mathbf{I} - \mathbf{A})^{-1} = \begin{pmatrix} \dfrac{s+1+1}{(s+1)^2+1} & \dfrac{1}{(s+1)^2+1} \\[2ex] \dfrac{-2}{(s+1)^2+1} & \dfrac{s+1-1}{(s+1)^2+1} \end{pmatrix}$$

and

$$e^{\mathbf{A}t} = \begin{pmatrix} e^{-t}\cos t + e^{-t}\sin t & e^{-t}\sin t \\ -2e^{-t}\sin t & e^{-t}\cos t - e^{-t}\sin t \end{pmatrix}.$$

The general solution of the system is then

$$\mathbf{X} = e^{\mathbf{A}t}\mathbf{C} = \begin{pmatrix} e^{-t}\cos t + e^{-t}\sin t & e^{-t}\sin t \\ -2e^{-t}\sin t & e^{-t}\cos t - e^{-t}\sin t \end{pmatrix}\begin{pmatrix} c_1 \\ c_2 \end{pmatrix}$$

$$= c_1\begin{pmatrix} 1 \\ 0 \end{pmatrix}e^{-t}\cos t + c_1\begin{pmatrix} 1 \\ -2 \end{pmatrix}e^{-t}\sin t + c_2\begin{pmatrix} 0 \\ 1 \end{pmatrix}e^{-t}\cos t + c_2\begin{pmatrix} 1 \\ -1 \end{pmatrix}e^{-t}\sin t$$

$$= c_1\begin{pmatrix} \cos t + \sin t \\ -2\sin t \end{pmatrix}e^{-t} + c_2\begin{pmatrix} \sin t \\ \cos t - \sin t \end{pmatrix}e^{-t}.$$

21. From equation (3) in the text

$$e^{t\mathbf{A}} = e^{t\mathbf{PDP}^{-1}} = \mathbf{I} + t(\mathbf{PDP}^{-1}) + \frac{1}{2!}t^2(\mathbf{PDP}^{-1})^2 + \frac{1}{3!}t^3(\mathbf{PDP}^{-1})^3 + \cdots$$

$$= \mathbf{P}\left[\mathbf{I} + t\mathbf{D} + \frac{1}{2!}(t\mathbf{D})^2 + \frac{1}{3!}(t\mathbf{D})^3 + \cdots\right]\mathbf{P}^{-1} = \mathbf{P}e^{t\mathbf{D}}\mathbf{P}^{-1}.$$

24. In Problem 20 it is shown that when

$$\mathbf{A} = \begin{pmatrix} 2 & 1 \\ 1 & 2 \end{pmatrix}$$

then

$$\mathbf{PDP}^{-1} = \begin{pmatrix} 2 & 1 \\ 1 & 2 \end{pmatrix}.$$

From Problems 20-22 and equation (1) in the text

$$\mathbf{X} = e^{t\mathbf{A}}\mathbf{C} = \mathbf{P}e^{t\mathbf{D}}\mathbf{P}^{-1}\mathbf{C}$$

$$= \begin{pmatrix} -e^t & e^{3t} \\ e^t & e^{3t} \end{pmatrix}\begin{pmatrix} e^t & 0 \\ 0 & e^{3t} \end{pmatrix}\begin{pmatrix} -\frac{1}{2}e^{-t} & \frac{1}{2}e^{-t} \\ \frac{1}{2}e^{3t} & \frac{1}{2}e^{-3t} \end{pmatrix}\begin{pmatrix} c_1 \\ c_2 \end{pmatrix}$$

$$= \begin{pmatrix} \frac{1}{2}e^t + \frac{1}{2}e^{9t} & -\frac{1}{2}e^t + \frac{1}{2}e^{3t} \\ -\frac{1}{2}e^t + \frac{1}{2}e^{9t} & \frac{1}{2}e^t + \frac{1}{2}e^{3t} \end{pmatrix}\begin{pmatrix} c_1 \\ c_2 \end{pmatrix}.$$

Chapter 8 in Review   *Hints, Suggestions, Solutions, and Examples*

3. Since

$$\begin{pmatrix} 4 & 6 & 6 \\ 1 & 3 & 2 \\ -1 & -4 & -3 \end{pmatrix}\begin{pmatrix} 3 \\ 1 \\ -1 \end{pmatrix} = \begin{pmatrix} 12 \\ 4 \\ -4 \end{pmatrix} = 4\begin{pmatrix} 3 \\ 1 \\ -1 \end{pmatrix},$$

we see that $\lambda = 4$ is an eigenvalue with eigenvector $\mathbf{K}_3$. The corresponding solution is $\mathbf{X}_3 = \mathbf{K}_3 e^{4t}$.

6. We have $\det(\mathbf{A} - \lambda\mathbf{I}) = (\lambda + 6)(\lambda + 2) = 0$ so that

$$\mathbf{X} = c_1 \begin{pmatrix} 1 \\ -1 \end{pmatrix} e^{-6t} + c_2 \begin{pmatrix} 1 \\ 1 \end{pmatrix} e^{-2t}.$$

9. We have $\det(\mathbf{A} - \lambda\mathbf{I}) = -(\lambda - 2)(\lambda - 4)(\lambda + 3) = 0$ so that

$$\mathbf{X} = c_1 \begin{pmatrix} -2 \\ 3 \\ 1 \end{pmatrix} e^{2t} + c_2 \begin{pmatrix} 0 \\ 1 \\ 1 \end{pmatrix} e^{4t} + c_3 \begin{pmatrix} 7 \\ 12 \\ -16 \end{pmatrix} e^{-3t}.$$

12. We have

$$\mathbf{X}_c = c_1 \begin{pmatrix} 2\cos t \\ -\sin t \end{pmatrix} e^t + c_2 \begin{pmatrix} 2\sin t \\ \cos t \end{pmatrix} e^t.$$

Then

$$\boldsymbol{\Phi} = \begin{pmatrix} 2\cos t & 2\sin t \\ -\sin t & \cos t \end{pmatrix} e^t, \quad \boldsymbol{\Phi}^{-1} = \begin{pmatrix} \frac{1}{2}\cos t & -\sin t \\ \frac{1}{2}\sin t & \cos t \end{pmatrix} e^{-t},$$

and

$$\mathbf{U} = \int \boldsymbol{\Phi}^{-1}\mathbf{F}\,dt = \int \begin{pmatrix} \cos t - \sec t \\ \sin t \end{pmatrix} dt = \begin{pmatrix} \sin t - \ln|\sec t + \tan t| \\ -\cos t \end{pmatrix},$$

so that

$$\mathbf{X}_p = \boldsymbol{\Phi}\mathbf{U} = \begin{pmatrix} -2\cos t \ln|\sec t + \tan t| \\ -1 + \sin t \ln|\sec t + \tan t| \end{pmatrix} e^t.$$

15. (a) Letting

$$\mathbf{K} = \begin{pmatrix} k_1 \\ k_2 \\ k_3 \end{pmatrix}$$

we note that $(\mathbf{A} - 2\mathbf{I})\mathbf{K} = \mathbf{0}$ implies that $3k_1 + 3k_2 + 3k_3 = 0$, so $k_1 = -(k_2 + k_3)$. Choosing $k_2 = 0$, $k_3 = 1$ and then $k_2 = 1$, $k_3 = 0$ we get

$$\mathbf{K}_1 = \begin{pmatrix} -1 \\ 0 \\ 1 \end{pmatrix} \quad \text{and} \quad \mathbf{K}_2 = \begin{pmatrix} -1 \\ 1 \\ 0 \end{pmatrix},$$

respectively. Thus,

$$\mathbf{X}_1 = \begin{pmatrix} -1 \\ 0 \\ 1 \end{pmatrix} e^{2t} \quad \text{and} \quad \mathbf{X}_2 = \begin{pmatrix} -1 \\ 1 \\ 0 \end{pmatrix} e^{2t}$$

are two solutions.

(b) From $\det(\mathbf{A} - \lambda\mathbf{I}) = \lambda^2(3 - \lambda) = 0$ we see that $\lambda_1 = 3$, and 0 is an eigenvalue of multiplicity two. Letting

$$\mathbf{K} = \begin{pmatrix} k_1 \\ k_2 \\ k_3 \end{pmatrix},$$

as in part (a), we note that $(\mathbf{A} - 0\mathbf{I})\mathbf{K} = \mathbf{A}\mathbf{K} = \mathbf{0}$ implies that $k_1 + k_2 + k_3 = 0$, so $k_1 = -(k_2 + k_3)$. Choosing $k_2 = 0$, $k_3 = 1$, and then $k_2 = 1$, $k_3 = 0$ we get

$$\mathbf{K}_2 = \begin{pmatrix} -1 \\ 0 \\ 1 \end{pmatrix} \quad \text{and} \quad \mathbf{K}_3 = \begin{pmatrix} -1 \\ 1 \\ 0 \end{pmatrix},$$

respectively. Since the eigenvector corresponding to $\lambda_1 = 3$ is

$$\mathbf{K}_1 = \begin{pmatrix} 1 \\ 1 \\ 1 \end{pmatrix},$$

the general solution of the system is

$$\mathbf{X} = c_1 \begin{pmatrix} 1 \\ 1 \\ 1 \end{pmatrix} e^{3t} + c_2 \begin{pmatrix} -1 \\ 0 \\ 1 \end{pmatrix} + c_3 \begin{pmatrix} -1 \\ 1 \\ 0 \end{pmatrix}.$$

9

Numerical Solutions of Ordinary Differential Equations

Section 9.1

Euler Methods and Error Analysis

The terminology and concepts listed below provide an outline of the main ideas encountered in this section. These can be useful when preparing for a quiz or test.

Terminology and Concepts

- error analysis for numerical solutions of ordinary DEs

- round-off error

- local and global truncation errors

- bound for local truncation error

- Euler's method for approximating the solution of a first-order IVP

- improved Euler's method for approximating the solution of a first-order IVP

- a predictor-corrector method

The basic skills listed below summarize the more mechanical types of problems encountered in the exercise set for this section.

Basic Skills

- use the improved Euler's method to approximate the solution of a first-order IVP

- find a bound on the local truncation error when Euler's method or the improved Euler's method is used to approximate the solution of an IVP

195

Exercises 9.1   *Hints, Suggestions, Solutions, and Examples*

3. $h=0.1$ $h=0.05$

x_n	y_n
0.00	0.0000
0.10	0.1005
0.20	0.2030
0.30	0.3098
0.40	0.4234
0.50	0.5470

x_n	y_n
0.00	0.0000
0.05	0.0501
0.10	0.1004
0.15	0.1512
0.20	0.2028
0.25	0.2554
0.30	0.3095
0.35	0.3652
0.40	0.4230
0.45	0.4832
0.50	0.5465

6. $h=0.1$ $h=0.05$

x_n	y_n
0.00	0.0000
0.10	0.0050
0.20	0.0200
0.30	0.0451
0.40	0.0805
0.50	0.1266

x_n	y_n
0.00	0.0000
0.05	0.0013
0.10	0.0050
0.15	0.0113
0.20	0.0200
0.25	0.0313
0.30	0.0451
0.35	0.0615
0.40	0.0805
0.45	0.1022
0.50	0.1266

9. $h=0.1$ $h=0.05$

x_n	y_n
1.00	1.0000
1.10	1.0095
1.20	1.0404
1.30	1.0967
1.40	1.1866
1.50	1.3260

x_n	y_n
1.00	1.0000
1.05	1.0024
1.10	1.0100
1.15	1.0228
1.20	1.0414
1.25	1.0663
1.30	1.0984
1.35	1.1389
1.40	1.1895
1.45	1.2526
1.50	1.3315

11. (*Hint*) Use the substitution $u = x + y - 1$ to find the analytic solution of the DE.

12. (a)

(b)

x_n	Euler	Imp. Euler
1.00	1.0000	1.0000
1.10	1.2000	1.2469
1.20	1.4938	1.6430
1.30	1.9711	2.4042
1.40	2.9060	4.5085

15. (a) Using Euler's method we obtain $y(0.1) \approx y_1 = 0.8$.

(b) Using $y'' = 5e^{-2x}$ we see that the local truncation error is

$$5e^{-2c}\frac{(0.1)^2}{2} = 0.025e^{-2c}.$$

Since e^{-2x} is a decreasing function, $e^{-2c} \leq e^0 = 1$ for $0 \leq c \leq 0.1$. Thus an upper bound for the local truncation error is $0.025(1) = 0.025$.

(c) Since $y(0.1) = 0.8234$, the actual error is $y(0.1) - y_1 = 0.0234$, which is less than 0.025.

(d) Using Euler's method with $h = 0.05$ we obtain $y(0.1) \approx y_2 = 0.8125$.

(e) The error in (d) is $0.8234 - 0.8125 = 0.0109$. With global truncation error $O(h)$, when the step size is halved we expect the error for $h = 0.05$ to be one-half the error when $h = 0.1$. Comparing 0.0109 with 0.0234 we see that this is the case.

18. (a) Using $y''' = -114e^{-3(x-1)}$ we see that the local truncation error is

$$\left| y'''(c) \frac{h^3}{6} \right| = 114e^{-3(x-1)} \frac{h^3}{6} = 19h^3 e^{-3(c-1)}.$$

(b) Since $e^{-3(x-1)}$ is a decreasing function for $1 \leq x \leq 1.5$, $e^{-3(c-1)} \leq e^{-3(1-1)} = 1$ for $1 \leq c \leq 1.5$ and

$$\left| y'''(c) \frac{h^3}{6} \right| \leq 19(0.1)^3(1) = 0.019.$$

(c) Using the improved Euler's method with $h = 0.1$ we obtain $y(1.5) \approx 2.080108$. With $h = 0.05$ we obtain $y(1.5) \approx 2.059166$.

(d) Since $y(1.5) = 2.053216$, the error for $h = 0.1$ is $E_{0.1} = 0.026892$, while the error for $h = 0.05$ is $E_{0.05} = 0.005950$. With global truncation error $O(h^2)$ we expect $E_{0.1}/E_{0.05} \approx 4$. We actually have $E_{0.1}/E_{0.05} = 4.52$.

Section 9.2

Runge-Kutta Methods

The terminology and concepts listed below provide an outline of the main ideas encountered in this section. These can be useful when preparing for a quiz or test.

Terminology and Concepts

- Runge-Kutta methods

- fourth-order Runge-Kutta method (RK4 method)

- truncation errors for the RK4 method

The basic skills listed below summarize the more mechanical types of problems encountered in the exercise set for this section.

Basic Skills

- use the RK4 method to approximate the solution of a first-order initial-value problem

Exercises 9.2   *Hints, Suggestions, Solutions, and Examples*

3.

x_n	y_n
1.00	5.0000
1.10	3.9724
1.20	3.2284
1.30	2.6945
1.40	2.3163
1.50	2.0533

6.

x_n	y_n
0.00	1.0000
0.10	1.1115
0.20	1.2530
0.30	1.4397
0.40	1.6961
0.50	2.0670

9.

x_n	y_n
0.00	0.5000
0.10	0.5213
0.20	0.5358
0.30	0.5443
0.40	0.5482
0.50	0.5493

12.

x_n	y_n
0.00	0.5000
0.10	0.5250
0.20	0.5498
0.30	0.5744
0.40	0.5987
0.50	0.6225

15. (a)

x_n	$h=0.05$	$h=0.1$
1.00	1.0000	1.0000
1.05	1.1112	
1.10	1.2511	1.2511
1.15	1.4348	
1.20	1.6934	1.6934
1.25	2.1047	
1.30	2.9560	2.9425
1.35	7.8981	
1.40	1.0608×10^{15}	903.0282

(b)

18. (a) Using $y^{(5)} = -1026e^{-3(x-1)}$ we see that the local truncation error is

$$\left| y^{(5)}(c)\,\frac{h^5}{120} \right| = 8.55h^5 e^{-3(c-1)}.$$

(b) Since $e^{-3(x-1)}$ is a decreasing function for $1 \le x \le 1.5$, $e^{-3(c-1)} \le e^{-3(1-1)} = 1$ for $1 \le c \le 1.5$ and

$$y^{(5)}(c)\,\frac{h^5}{120} \le 8.55(0.1)^5(1) = 0.0000855.$$

(c) Using the RK4 method with $h = 0.1$ we obtain $y(1.5) \approx 2.053338827$. With $h = 0.05$ we obtain $y(1.5) \approx 2.053222989$.

Section 9.3

Multistep Methods

The terminology and concepts listed below provide an outline of the main ideas encountered in this section. These can be useful when preparing for a quiz or test.

Terminology and Concepts

- predictor-corrector methods
- Adams-Bashforth-Moulton method

The basic skills listed below summarize the more mechanical types of problems encountered in the exercise set for this section.

Basic Skills

- use the Adams-Bashforth-Moulton predictor-corrector method to approximate the solution of a first-order IVP

Exercises 9.3 *Hints, Suggestions, Solutions, and Examples*

In the tables in this section "ABM" stands for "Adams-Bashforth-Moulton."

3. The first predictor is $y_4^* = 0.73318477$.

x_n	y_n	
0.0	1.00000000	init. cond.
0.2	0.73280000	RK4
0.4	0.64608032	RK4
0.6	0.65851653	RK4
0.8	0.72319464	ABM

6. The first predictor for $h = 0.2$ is $y_4^* = 3.34828434$.

x_n	h=0.2		h=0.1	
0.0	1.00000000	init. cond.	1.00000000	init. cond.
0.1			1.21017082	RK4
0.2	1.44139950	RK4	1.44140511	RK4
0.3			1.69487942	RK4
0.4	1.97190167	RK4	1.97191536	ABM
0.5			2.27400341	ABM
0.6	2.60280694	RK4	2.60283209	ABM
0.7			2.96031780	ABM
0.8	3.34860927	ABM	3.34863769	ABM
0.9			3.77026548	ABM
1.0	4.22797875	ABM	4.22801028	ABM

Section 9.4   Higher-Order Equations and Systems

The terminology and concepts listed below provide an outline of the main ideas encountered in this section. These can be useful when preparing for a quiz or test.

Terminology and Concepts

- express a second-order IVP as an IVP for a first-order system
- express a higher-order IVP as an IVP for a first-order system
- numerical solution of an IVP for a first-order system using Euler's method and the RK4 method

The basic skills listed below summarize the more mechanical types of problems encountered in the exercise set for this section.

Basic Skills

- solve an IVP for a first-order system consisting of two equations

Exercises 9.4   *Hints, Suggestions, Solutions, and Examples*

3. The substitution $y' = u$ leads to the system

$$y' = u, \qquad u' = 4u - 4y.$$

Using formula (4) in the text with x corresponding to t, y corresponding to x, and u corresponding to y, we obtain the table shown.

x_n	h=0.2 y_n	h=0.2 u_n	h=0.1 y_n	h=0.1 u_n
0.0	-2.0000	1.0000	-2.0000	1.0000
0.1			-1.8321	2.4427
0.2	-1.4928	4.4731	-1.4919	4.4753

6. Using $h = 0.1$, the RK4 method for a system, and a numerical solver, we obtain

t_n	h=0.2 i_{1n}	h=0.2 i_{3n}
0.0	0.0000	0.0000
0.1	2.5000	3.7500
0.2	2.8125	5.7813
0.3	2.0703	7.4023
0.4	0.6104	9.1919
0.5	-1.5619	11.4877

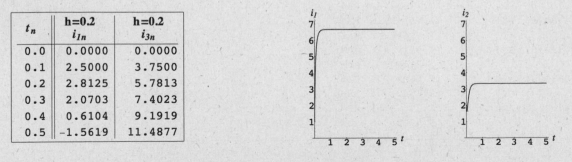

9.

t_n	h=0.2 x_n	h=0.2 y_n	h=0.1 x_n	h=0.1 y_n
0.0	-3.0000	5.0000	-3.0000	5.0000
0.1			-3.4790	4.6707
0.2	-3.9123	4.2857	-3.9123	4.2857

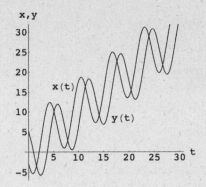

12. Solving for x' and y' we obtain the system

$$x' = \frac{1}{2}y - 3t^2 + 2t - 5$$

$$y' = -\frac{1}{2}y + 3t^2 + 2t + 5.$$

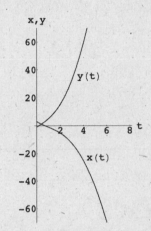

t_n	h=0.2 x_n	h=0.2 y_n	h=0.1 x_n	h=0.1 y_n
0.0	3.0000	-1.0000	3.0000	-1.0000
0.1			2.4727	-0.4527
0.2	1.9867	0.0933	1.9867	0.0933

Section 9.5

Second-Order Boundary-Value Problems

The terminology and concepts listed below provide an outline of the main ideas encountered in this section. These can be useful when preparing for a quiz or test.

Terminology and Concepts

- difference quotient
- finite difference
- forward difference
- backward difference
- central difference
- finite difference method

The basic skills listed below summarize the more mechanical types of problems encountered in the exercise set for this section.

Basic Skills

• use the finite difference method to approximate the solution of a second-order boundary-value problem

Exercises 9.5   *Hints, Suggestions, Solutions, and Examples*

3. We identify $P(x) = 2$, $Q(x) = 1$, $f(x) = 5x$, and $h = (1 - 0)/5 = 0.2$. Then the finite difference equation is

$$1.2y_{i+1} - 1.96y_i + 0.8y_{i-1} = 0.04(5x_i).$$

The solution of the corresponding linear system gives

x	0.0	0.2	0.4	0.6	0.8	1.0
y	0.0000	-0.2259	-0.3356	-0.3308	-0.2167	0.0000

6. We identify $P(x) = 5$, $Q(x) = 0$, $f(x) = 4\sqrt{x}$, and $h = (2 - 1)/6 = 0.1667$. Then the finite difference equation is

$$1.4167y_{i+1} - 2y_i + 0.5833y_{i-1} = 0.2778(4\sqrt{x_i}).$$

The solution of the corresponding linear system gives

x	1.0000	1.1667	1.3333	1.5000	1.6667	1.8333	2.0000
y	1.0000	-0.5918	-1.1626	-1.3070	-1.2704	-1.1541	-1.0000

9. We identify $P(x) = 1 - x$, $Q(x) = x$, $f(x) = x$, and $h = (1 - 0)/10 = 0.1$. Then the finite difference equation is

$$[1 + 0.05(1 - x_i)]y_{i+1} + [-2 + 0.01x_i]y_i + [1 - 0.05(1 - x_i)]y_{i-1} = 0.01x_i.$$

The solution of the corresponding linear system gives

x	0.0	0.1	0.2	0.3	0.4	0.5	0.6
y	0.0000	0.2660	0.5097	0.7357	0.9471	1.1465	1.3353

0.7	0.8	0.9	1.0
1.5149	1.6855	1.8474	2.0000

12. We identify $P(r) = 2/r$, $Q(r) = 0$, $f(r) = 0$, and $h = (4 - 1)/6 = 0.5$. Then the finite difference equation is

$$\left(1 + \frac{0.5}{r_i}\right)u_{i+1} - 2u_i + \left(1 - \frac{0.5}{r_i}\right)u_{i-1} = 0.$$

The solution of the corresponding linear system gives

r	1.0	1.5	2.0	2.5	3.0	3.5	4.0
u	50.0000	72.2222	83.3333	90.0000	94.4444	97.6190	100.0000

Chapter 9 in Review

Hints, Suggestions, Solutions, and Examples

3.

x_n	Euler h=0.1	Euler h=0.05	Imp. Euler h=0.1	Imp. Euler h=0.05	RK4 h=0.1	RK4 h=0.05
0.50	0.5000	0.5000	0.5000	0.5000	0.5000	0.5000
0.55		0.5500		0.5512		0.5512
0.60	0.6000	0.6024	0.6048	0.6049	0.6049	0.6049
0.65		0.6573		0.6609		0.6610
0.70	0.7095	0.7144	0.7191	0.7193	0.7194	0.7194
0.75		0.7739		0.7800		0.7801
0.80	0.8283	0.8356	0.8427	0.8430	0.8431	0.8431
0.85		0.8996		0.9082		0.9083
0.90	0.9559	0.9657	0.9752	0.9755	0.9757	0.9757
0.95		1.0340		1.0451		1.0452
1.00	1.0921	1.1044	1.1163	1.1168	1.1169	1.1169

6. The first predictor is $y_3^* = 1.14822731$.

x_n	y_n	
0.0	2.00000000	init. cond.
0.1	1.65620000	RK4
0.2	1.41097281	RK4
0.3	1.24645047	RK4
0.4	1.14796764	ABM

Appendices

Appendix I

Gamma Function

Solutions for Appendix I

3. If $t = x^3$, then $dt = 3x^2\,dx$ and $x^4\,dx = \frac{1}{3}t^{2/3}\,dt$. Now

$$\int_0^\infty x^4 e^{-x^3}\,dx = \int_0^\infty \frac{1}{3}t^{2/3}e^{-t}\,dt = \frac{1}{3}\int_0^\infty t^{2/3}e^{-t}\,dt$$

$$= \frac{1}{3}\Gamma\left(\frac{5}{3}\right) = \frac{1}{3}(0.89) \approx 0.297.$$

6. For $x > 0$ we integrate by parts:

$$\Gamma(x+1) = \int_0^\infty t^x e^{-t}\,dt$$

$$\boxed{\begin{array}{ll} u = t^x & dv = e^{-t}\,dt \\ du = xt^{x-1}\,dt & v = -e^{-t} \end{array}}$$

$$= -t^x e^{-t}\,\Big|_0^\infty - \int_0^\infty xt^{x-1}(-e^{-t})\,dt$$

$$= x\int_0^\infty t^{x-1}e^{-t}\,dt$$

$$= x\Gamma(x).$$

Appendix II Matrices

Solutions for Appendix II

3. (a) $\mathbf{AB} = \begin{pmatrix} -2-9 & 12-6 \\ 5+12 & -30+8 \end{pmatrix} = \begin{pmatrix} -11 & 6 \\ 17 & -22 \end{pmatrix}$

(b) $\mathbf{BA} = \begin{pmatrix} -2-30 & 3+24 \\ 6-10 & -9+8 \end{pmatrix} = \begin{pmatrix} -32 & 27 \\ -4 & -1 \end{pmatrix}$

(c) $\mathbf{A}^2 = \begin{pmatrix} 4+15 & -6-12 \\ -10-20 & 15+16 \end{pmatrix} = \begin{pmatrix} 19 & -18 \\ -30 & 31 \end{pmatrix}$

(d) $\mathbf{B}^2 = \begin{pmatrix} 1+18 & -6+12 \\ -3+6 & 18+4 \end{pmatrix} = \begin{pmatrix} 19 & 6 \\ 3 & 22 \end{pmatrix}$

6. (a) $\mathbf{AB} = \begin{pmatrix} 5 & -6 & 7 \end{pmatrix} \begin{pmatrix} 3 \\ 4 \\ -1 \end{pmatrix} = (-16)$

(b) $\mathbf{BA} = \begin{pmatrix} 3 \\ 4 \\ -1 \end{pmatrix} \begin{pmatrix} 5 & -6 & 7 \end{pmatrix} = \begin{pmatrix} 15 & -18 & 21 \\ 20 & -24 & 28 \\ -5 & 6 & -7 \end{pmatrix}$

(c) $(\mathbf{BA})\mathbf{C} = \begin{pmatrix} 15 & -18 & 21 \\ 20 & -24 & 28 \\ -5 & 6 & -7 \end{pmatrix} \begin{pmatrix} 1 & 2 & 4 \\ 0 & 1 & -1 \\ 3 & 2 & 1 \end{pmatrix} = \begin{pmatrix} 78 & 54 & 99 \\ 104 & 72 & 132 \\ -26 & -18 & -33 \end{pmatrix}$

(d) Since $\mathbf{AB}$ is 1×1 and $\mathbf{C}$ is 3×3 the product $(\mathbf{AB})\mathbf{C}$ is not defined.

9. (a) $(\mathbf{AB})^T = \begin{pmatrix} 7 & 10 \\ 38 & 75 \end{pmatrix}^T = \begin{pmatrix} 7 & 38 \\ 10 & 75 \end{pmatrix}$

(b) $\mathbf{B}^T\mathbf{A}^T = \begin{pmatrix} 5 & -2 \\ 10 & -5 \end{pmatrix} \begin{pmatrix} 3 & 8 \\ 4 & 1 \end{pmatrix} = \begin{pmatrix} 7 & 38 \\ 10 & 75 \end{pmatrix}$

12. $\begin{pmatrix} 6t \\ 3t^2 \\ -3t \end{pmatrix} + \begin{pmatrix} -t+1 \\ -t^2+t \\ 3t-3 \end{pmatrix} - \begin{pmatrix} 6t \\ 8 \\ -10t \end{pmatrix} = \begin{pmatrix} -t+1 \\ 2t^2+t-8 \\ 10t-3 \end{pmatrix}$

15. Since $\det \mathbf{A} = 0$, $\mathbf{A}$ is singular.

18. Since $\det \mathbf{A} = -6$, $\mathbf{A}$ is nonsingular.

$$\mathbf{A}^{-1} = -\frac{1}{6}\begin{pmatrix} 2 & -10 \\ -2 & 7 \end{pmatrix}$$

21. Since $\det \mathbf{A} = -9$, $\mathbf{A}$ is nonsingular. The cofactors are

$$\begin{array}{ccc} A_{11} = -2 & A_{12} = -13 & A_{13} = 8 \\ A_{21} = -2 & A_{22} = 5 & A_{23} = -1 \\ A_{31} = -1 & A_{32} = 7 & A_{33} = -5. \end{array}$$

Then

$$\mathbf{A}^{-1} = -\frac{1}{9}\begin{pmatrix} -2 & -13 & 8 \\ -2 & 5 & -1 \\ -1 & 7 & -5 \end{pmatrix}^T = -\frac{1}{9}\begin{pmatrix} -2 & -2 & -1 \\ -13 & 5 & 7 \\ 8 & -1 & -5 \end{pmatrix}.$$

24. Since $\det \mathbf{A}(t) = 2e^{2t} \neq 0$, $\mathbf{A}$ is nonsingular.

$$\mathbf{A}^{-1} = \frac{1}{2}e^{-2t}\begin{pmatrix} e^t \sin t & 2e^t \cos t \\ -e^t \cos t & 2e^t \sin t \end{pmatrix}$$

27. $\mathbf{X} = \begin{pmatrix} 2e^{2t} + 8e^{-3t} \\ -2e^{2t} + 4e^{-3t} \end{pmatrix}$ so that $\dfrac{d\mathbf{X}}{dt} = \begin{pmatrix} 4e^{2t} - 24e^{-3t} \\ -4e^{2t} - 12e^{-3t} \end{pmatrix}.$

30. (a) $\dfrac{d\mathbf{A}}{dt} = \begin{pmatrix} -2t/(t^2+1)^2 & 3 \\ 2t & 1 \end{pmatrix}$

(b) $\dfrac{d\mathbf{B}}{dt} = \begin{pmatrix} 6 & 0 \\ -1/t^2 & 4 \end{pmatrix}$

(c) $\displaystyle\int_0^1 \mathbf{A}(t)\,dt = \begin{pmatrix} \tan^{-1} t & \frac{3}{2}t^2 \\ \frac{1}{3}t^3 & \frac{1}{2}t^2 \end{pmatrix}\Big|_{t=0}^{t=1} = \begin{pmatrix} \frac{\pi}{4} & \frac{3}{2} \\ \frac{1}{3} & \frac{1}{2} \end{pmatrix}$

(d) $\displaystyle\int_1^2 \mathbf{B}(t)\,dt = \begin{pmatrix} 3t^2 & 2t \\ \ln t & 2t^2 \end{pmatrix}\Big|_{t=1}^{t=2} = \begin{pmatrix} 9 & 2 \\ \ln 2 & 6 \end{pmatrix}$

(e) $\mathbf{A}(t)\mathbf{B}(t) = \begin{pmatrix} 6t/(t^2+1)+3 & 2/(t^2+1)+12t^2 \\ 6t^3+1 & 2t^2+4t^2 \end{pmatrix}$

(f) $\dfrac{d}{dt}\mathbf{A}(t)\mathbf{B}(t) = \begin{pmatrix} (6-6t^2)/(t^2+1)^2 & -4t/(t^2+1)^2 + 24t \\ 18t^2 & 12t \end{pmatrix}$

(g) $\displaystyle\int_1^t \mathbf{A}(s)\mathbf{B}(s)\,ds = \int_1^t \begin{pmatrix} 6s/(s^2+1)+3 & 2/(s^2+1)+12s^2 \\ 6s^3+1 & 6s^2 \end{pmatrix} ds$

$$= \begin{pmatrix} 3s+3\ln(s^2+1) & 4s^3+2\tan^{-1}s \\ \frac{3}{2}s^4 + s & 2s^3 \end{pmatrix}\Bigg|_1^t$$

$$= \begin{pmatrix} 3t+3\ln(t^2+1)-3-3\ln 2 & 4t^3+2\tan^{-1}t-4-\pi/2 \\ \frac{3}{2}t^4+t-\frac{5}{2} & 2t^3-2 \end{pmatrix}$$

33. $\begin{pmatrix} 1 & -1 & -5 & | & 7 \\ 5 & 4 & -16 & | & -10 \\ 0 & 1 & 1 & | & -5 \end{pmatrix} \Longrightarrow \begin{pmatrix} 1 & -1 & -5 & | & 7 \\ 0 & 1 & 1 & | & -5 \\ 0 & 9 & 9 & | & -45 \end{pmatrix} \Longrightarrow \begin{pmatrix} 1 & 0 & -4 & | & 2 \\ 0 & 1 & 1 & | & -5 \\ 0 & 0 & 0 & | & 0 \end{pmatrix}$

Letting $z = t$ we find $y = -5-t$, and $x = 2+4t$.

36. $\begin{pmatrix} 1 & 0 & 2 & | & 8 \\ 1 & 2 & -2 & | & 4 \\ 2 & 5 & -6 & | & 6 \end{pmatrix} \Longrightarrow \begin{pmatrix} 1 & 0 & 2 & | & 8 \\ 0 & 2 & -4 & | & -4 \\ 0 & 5 & -10 & | & -10 \end{pmatrix} \Longrightarrow \begin{pmatrix} 1 & 0 & 2 & | & 8 \\ 0 & 1 & -2 & | & -2 \\ 0 & 0 & 0 & | & 0 \end{pmatrix}$

Letting $z = t$ we find $y = -2+2t$, and $x = 8-2t$.

39. $\begin{pmatrix} 1 & 2 & 4 & | & 2 \\ 2 & 4 & 3 & | & 1 \\ 1 & 2 & -1 & | & 7 \end{pmatrix} \Longrightarrow \begin{pmatrix} 1 & 2 & 4 & | & 2 \\ 0 & 0 & -5 & | & -3 \\ 0 & 0 & -5 & | & 5 \end{pmatrix} \Longrightarrow \begin{pmatrix} 1 & 2 & 0 & | & -2/5 \\ 0 & 0 & 1 & | & 3/5 \\ 0 & 0 & 0 & | & 8 \end{pmatrix}$

There is no solution.

42. $\begin{pmatrix} 2 & 4 & -2 & | & 1 & 0 & 0 \\ 4 & 2 & -2 & | & 0 & 1 & 0 \\ 8 & 10 & -6 & | & 0 & 0 & 1 \end{pmatrix} \xrightarrow[\text{operations}]{\text{row}} \begin{pmatrix} 1 & 2 & -1 & | & \frac{1}{2} & 0 & 0 \\ 0 & 1 & -\frac{1}{3} & | & \frac{1}{3} & -\frac{1}{6} & 0 \\ 0 & 0 & 0 & | & -2 & -1 & 1 \end{pmatrix}$; **A** is singular.

45. $\begin{pmatrix} 1 & 2 & 3 & 1 & | & 1 & 0 & 0 & 0 \\ -1 & 0 & 2 & 1 & | & 0 & 1 & 0 & 0 \\ 2 & 1 & -3 & 0 & | & 0 & 0 & 1 & 0 \\ 1 & 1 & 2 & 1 & | & 0 & 0 & 0 & 1 \end{pmatrix} \xrightarrow[\text{operations}]{\text{row}} \begin{pmatrix} 1 & 2 & 3 & 1 & | & 1 & 0 & 0 & 0 \\ 0 & 1 & \frac{5}{2} & 1 & | & \frac{1}{2} & \frac{1}{2} & 0 & 0 \\ 0 & 0 & 1 & -\frac{2}{3} & | & \frac{1}{3} & -1 & -\frac{2}{3} & 0 \\ 0 & 0 & 0 & 1 & | & -\frac{1}{2} & 1 & \frac{1}{2} & \frac{1}{2} \end{pmatrix}$

$$\xrightarrow[\text{operations}]{\text{row}} \begin{pmatrix} 1 & 0 & 0 & 0 & \bigg| & -\frac{1}{2} & -\frac{2}{3} & -\frac{1}{6} & \frac{7}{6} \\ 0 & 1 & 0 & 0 & \bigg| & 1 & \frac{1}{3} & \frac{1}{3} & -\frac{4}{3} \\ 0 & 0 & 1 & 0 & \bigg| & 0 & -\frac{1}{3} & -\frac{1}{3} & \frac{1}{3} \\ 0 & 0 & 0 & 1 & \bigg| & -\frac{1}{2} & 1 & \frac{1}{2} & \frac{1}{2} \end{pmatrix}; \quad \mathbf{A}^{-1} = \begin{pmatrix} -\frac{1}{2} & -\frac{2}{3} & -\frac{1}{6} & \frac{7}{6} \\ 1 & \frac{1}{3} & \frac{1}{3} & -\frac{4}{3} \\ 0 & -\frac{1}{3} & -\frac{1}{3} & \frac{1}{3} \\ -\frac{1}{2} & 1 & \frac{1}{2} & \frac{1}{2} \end{pmatrix}$$

48. We solve

$$\det(\mathbf{A} - \lambda \mathbf{I}) = \begin{vmatrix} 2 - \lambda & 1 \\ 2 & 1 - \lambda \end{vmatrix} = \lambda(\lambda - 3) = 0.$$

For $\lambda_1 = 0$ we have

$$\begin{pmatrix} 2 & 1 & \big| & 0 \\ 2 & 1 & \big| & 0 \end{pmatrix} \Longrightarrow \begin{pmatrix} 1 & 1/2 & \big| & 0 \\ 0 & 0 & \big| & 0 \end{pmatrix}$$

so that $k_1 = -\frac{1}{2}k_2$. If $k_2 = 2$ then

$$\mathbf{K}_1 = \begin{pmatrix} -1 \\ 2 \end{pmatrix}.$$

For $\lambda_2 = 3$ we have

$$\begin{pmatrix} -1 & 1 & \big| & 0 \\ 2 & -2 & \big| & 0 \end{pmatrix} \Longrightarrow \begin{pmatrix} 1 & -1 & \big| & 0 \\ 0 & 0 & \big| & 0 \end{pmatrix}$$

so that $k_1 = k_2$. If $k_2 = 1$ then

$$\mathbf{K}_2 = \begin{pmatrix} 1 \\ 1 \end{pmatrix}.$$

51. We solve

$$\det(\mathbf{A} - \lambda \mathbf{I}) = \begin{vmatrix} 5 - \lambda & -1 & 0 \\ 0 & -5 - \lambda & 9 \\ 5 & -1 & -\lambda \end{vmatrix} = \lambda(4 - \lambda)(\lambda + 4) = 0.$$

If $\lambda_1 = 0$ then

$$\begin{pmatrix} 5 & -1 & 0 & \big| & 0 \\ 0 & -5 & 9 & \big| & 0 \\ 5 & -1 & 0 & \big| & 0 \end{pmatrix} \Longrightarrow \begin{pmatrix} 1 & 0 & -9/25 & \big| & 0 \\ 0 & 1 & -9/5 & \big| & 0 \\ 0 & 0 & 0 & \big| & 0 \end{pmatrix}$$

so that $k_1 = \frac{9}{25}k_3$ and $k_2 = \frac{9}{5}k_3$. If $k_3 = 25$ then

$$\mathbf{K}_1 = \begin{pmatrix} 9 \\ 45 \\ 25 \end{pmatrix}.$$

If $\lambda_2 = 4$ then

$$\begin{pmatrix} 1 & -1 & 0 & | & 0 \\ 0 & -9 & 9 & | & 0 \\ 5 & -1 & -4 & | & 0 \end{pmatrix} \implies \begin{pmatrix} 1 & 0 & -1 & | & 0 \\ 0 & 1 & -1 & | & 0 \\ 0 & 0 & 0 & | & 0 \end{pmatrix}$$

so that $k_1 = k_3$ and $k_2 = k_3$. If $k_3 = 1$ then

$$\mathbf{K}_2 = \begin{pmatrix} 1 \\ 1 \\ 1 \end{pmatrix}.$$

If $\lambda_3 = -4$ then

$$\begin{pmatrix} 9 & -1 & 0 & | & 0 \\ 0 & -1 & 9 & | & 0 \\ 5 & -1 & 4 & | & 0 \end{pmatrix} \implies \begin{pmatrix} 1 & 0 & -1 & | & 0 \\ 0 & 1 & -9 & | & 0 \\ 0 & 0 & 0 & | & 0 \end{pmatrix}$$

so that $k_1 = k_3$ and $k_2 = 9k_3$. If $k_3 = 1$ then

$$\mathbf{K}_3 = \begin{pmatrix} 1 \\ 9 \\ 1 \end{pmatrix}.$$

54. We solve

$$\det(\mathbf{A} - \lambda\mathbf{I}) = \begin{vmatrix} 1-\lambda & 6 & 0 \\ 0 & 2-\lambda & 1 \\ 0 & 1 & 2-\lambda \end{vmatrix} = (3-\lambda)(1-\lambda)^2 = 0.$$

For $\lambda = 3$ we have

$$\begin{pmatrix} -2 & 6 & 0 & | & 0 \\ 0 & 0 & 0 & | & 0 \\ 0 & 1 & -1 & | & 0 \end{pmatrix} \implies \begin{pmatrix} 1 & 0 & -3 & | & 0 \\ 0 & 1 & -1 & | & 0 \\ 0 & 0 & 0 & | & 0 \end{pmatrix}$$

so that $k_1 = 3k_3$ and $k_2 = k_3$. If $k_3 = 1$ then

$$\mathbf{K}_1 = \begin{pmatrix} 3 \\ 1 \\ 1 \end{pmatrix}.$$

For $\lambda_2 = \lambda_3 = 1$ we have

$$\begin{pmatrix} 0 & 6 & 0 & | & 0 \\ 0 & 1 & 1 & | & 0 \\ 0 & 1 & 1 & | & 0 \end{pmatrix} \implies \begin{pmatrix} 0 & 1 & 0 & | & 0 \\ 0 & 0 & 1 & | & 0 \\ 0 & 0 & 0 & | & 0 \end{pmatrix}$$

so that $k_2 = 0$ and $k_3 = 0$. If $k_1 = 1$ then

$$\mathbf{K}_2 = \begin{pmatrix} 1 \\ 0 \\ 0 \end{pmatrix}.$$

57. Let

$$\mathbf{A} = \begin{pmatrix} a_{11} & a_{12} \\ a_{21} & a_{22} \end{pmatrix}.$$

Then

$$\frac{d}{dt}[\mathbf{A}(t)\mathbf{X}(t)] = \frac{d}{dt}\begin{pmatrix} a_1 & a_2 \\ a_3 & a_4 \end{pmatrix}\begin{pmatrix} x_1 \\ x_2 \end{pmatrix} = \frac{d}{dt}\begin{pmatrix} a_1 x_1 + a_2 x_2 \\ a_3 x_1 + a_4 x_2 \end{pmatrix} = \begin{pmatrix} a_1 x_1' + a_1' x_1 + a_2 x_2' + a_2' x_2 \\ a_3 x_1' + a_3' x_1 + a_4 x_2' + a_4' x_2 \end{pmatrix}$$

$$= \begin{pmatrix} a_1 & a_2 \\ a_3 & a_4 \end{pmatrix}\begin{pmatrix} x_1' \\ x_2' \end{pmatrix} + \begin{pmatrix} a_1' & a_2' \\ a_3' & a_4' \end{pmatrix}\begin{pmatrix} x_1 \\ x_2 \end{pmatrix} = \mathbf{A}(t)\mathbf{X}'(t) + \mathbf{A}'(t)\mathbf{X}(t).$$

60. Since

$$(\mathbf{AB})(\mathbf{B}^{-1}\mathbf{A}^{-1}) = \mathbf{A}(\mathbf{BB}^{-1})\mathbf{A}^{-1} = \mathbf{AIA}^{-1} = \mathbf{AA}^{-1} = \mathbf{I}$$

and

$$(\mathbf{B}^{-1}\mathbf{A}^{-1})(\mathbf{AB}) = \mathbf{B}^{-1}(\mathbf{A}^{-1}\mathbf{A})\mathbf{B} = \mathbf{B}^{-1}\mathbf{IB} = \mathbf{B}^{-1}\mathbf{B} = \mathbf{I}$$

we have

$$(\mathbf{AB})^{-1} = \mathbf{B}^{-1}\mathbf{A}^{-1}.$$

Table of Integrals

1. $\displaystyle\int u\,dv = uv - \int v\,du$

2. $\displaystyle\int u^n\,du = \frac{1}{n+1}u^{n+1} + C, \quad n \neq -1$

3. $\displaystyle\int \frac{du}{u} = \ln|u| + C$

4. $\displaystyle\int e^u\,du = e^u + C$

5. $\displaystyle\int a^u\,du = \frac{1}{\ln a}a^u + C$

6. $\displaystyle\int \sin u\,du = -\cos u + C$

7. $\displaystyle\int \cos u\,du = \sin u + C$

8. $\displaystyle\int \sec^2 u\,du = \tan u + C$

9. $\displaystyle\int \csc^2 u\,du = -\cot u + C$

10. $\displaystyle\int \sec u\tan u\,du = \sec u + C$

11. $\displaystyle\int \csc u\cot u\,du = -\csc u + C$

12. $\displaystyle\int \tan u\,du = -\ln|\cos u| + C$

13. $\displaystyle\int \cot u\,du = \ln|\sin u| + C$

14. $\displaystyle\int \sec u\,du = \ln|\sec u + \tan u| + C$

15. $\displaystyle\int \csc u\,du = \ln|\csc u - \cot u| + C$

16. $\displaystyle\int \frac{du}{\sqrt{a^2 - u^2}} = \sin^{-1}\frac{u}{a} + C$

17. $\displaystyle\int \frac{du}{a^2 + u^2} = \frac{1}{a}\tan^{-1}\frac{u}{a} + C$

18. $\displaystyle\int \frac{du}{u\sqrt{u^2 - a^2}} = \frac{1}{a}\sec^{-1}\frac{u}{a} + C$

19. $\displaystyle\int \frac{du}{a^2 - u^2} = \frac{1}{2a}\ln\left|\frac{u+a}{u-a}\right| + C$

20. $\displaystyle\int \frac{du}{u^2 - a^2} = \frac{1}{2a}\ln\left|\frac{u-a}{u+a}\right| + C$

21. $\displaystyle\int \sin^2 u\,du = \frac{1}{2}u - \frac{1}{4}\sin 2u + C$

22. $\displaystyle\int \cos^2 u\,du = \frac{1}{2}u + \frac{1}{4}\sin 2u + C$

23. $\displaystyle\int \tan^2 u\,du = \tan u - u + C$

24. $\displaystyle\int \cot^2 u\,du = -\cot u - u + C$

25. $\displaystyle\int \sin^3 u\,du = -\frac{1}{3}(2 + \sin^2 u)\cos u + C$

26. $\displaystyle\int \cos^3 u\,du = \frac{1}{3}(2 + \cos^2 u)\sin u + C$

27. $\displaystyle\int \tan^3 u\,du = \frac{1}{2}\tan^2 u + \ln|\cos u| + C$

28. $\displaystyle\int \cot^3 u\,du = -\frac{1}{2}\cot^2 u - \ln|\sin u| + C$

29. $\displaystyle\int \sec^3 u\,du = \frac{1}{2}\sec u\tan u + \frac{1}{2}\ln|\sec u + \tan u| + C$

30. $\displaystyle\int \csc^3 u\,du = -\frac{1}{2}\csc u\cot u + \frac{1}{2}\ln|\csc u - \cot u| + C$

31. $\displaystyle\int \sin^n u\,du = -\frac{1}{n}\sin^{n-1} u\cos u + \frac{n-1}{n}\int \sin^{n-2} u\,du$

32. $\displaystyle\int \cos^n u\,du = \frac{1}{n}\cos^{n-1} u\sin u + \frac{n-1}{n}\int \cos^{n-2} u\,du$

33. $\int \tan^n u\, du = \dfrac{1}{n-1} \tan^{n-1} u - \int \tan^{n-2} u\, du$

34. $\int \cot^n u\, du = \dfrac{-1}{n-1} \cot^{n-1} u - \int \cot^{n-2} u\, du$

35. $\int \sec^n u\, du = \dfrac{1}{n-1} \tan u \sec^{n-2} u + \dfrac{n-2}{n-1} \int \sec^{n-2} u\, du$

36. $\int \csc^n u\, du = \dfrac{-1}{n-1} \cot u \csc^{n-2} u + \dfrac{n-2}{n-1} \int \csc^{n-2} u\, du$

37. $\int \sin au \sin bu\, du = \dfrac{\sin(a-b)u}{2(a-b)} - \dfrac{\sin(a+b)u}{2(a+b)} + C, \quad a \neq b$

38. $\int \cos au \cos bu\, du = \dfrac{\sin(a-b)u}{2(a-b)} + \dfrac{\sin(a+b)u}{2(a+b)} + C, \quad a \neq b$

39. $\int \sin au \cos bu\, du = -\dfrac{\cos(a-b)u}{2(a-b)} - \dfrac{\cos(a+b)u}{2(a+b)} + C, \quad a \neq b$

40. $\int u \sin u\, du = \sin u - u \cos u + C$

41. $\int u \cos u\, du = \cos u + u \sin u + C$

42. $\int u^n \sin u\, du = -u^n \cos u + n \int u^{n-1} \cos u\, du$

43. $\int u^n \cos u\, du = u^n \sin u - n \int u^{n-1} \sin u\, du$

44. $\int \sin^n u \cos^m u\, du = -\dfrac{\sin^{n-1} u \cos^{m+1} u}{n+m} + \dfrac{n-1}{n+m} \int \sin^{n-2} u \cos^m u\, du$

$\qquad = \dfrac{\sin^{n+1} u \cos^{m-1} u}{n+m} + \dfrac{m-1}{n+m} \int \sin^n u \cos^{m-2} u\, du$

45. $\int \sin^{-1} u\, du = u \sin^{-1} u + \sqrt{1-u^2} + C$

46. $\int \cos^{-1} u\, du = u \cos^{-1} u + \sqrt{1-u^2} + C$

47. $\int \tan^{-1} u\, du = u \tan^{-1} u - \dfrac{1}{2} \ln(1+u^2) + C$

48. $\int u \sin^{-1} u\, du = \dfrac{2u^2-1}{4} \sin^{-1} u + \dfrac{u\sqrt{1-u^2}}{4} + C$

49. $\int u \cos^{-1} u\, du = \dfrac{2u^2-1}{4} \cos^{-1} u - \dfrac{u\sqrt{1-u^2}}{4} + C$

50. $\int u \tan^{-1} u\, du = \dfrac{u^2+1}{2} \tan^{-1} u - \dfrac{u}{2} + C$

51. $\int u e^{au}\, du = \dfrac{1}{a^2}(au-1)e^{au} + C$

52. $\int u^n e^{au}\, du = \dfrac{1}{a} u^n e^{au} - \dfrac{n}{a} \int u^{n-1} e^{au}\, du$

53. $\int e^{au} \sin bu\, du = \dfrac{e^{au}}{a^2+b^2}(a \sin bu - b \cos bu) + C$

54. $\int e^{au} \cos bu\, du = \dfrac{e^{au}}{a^2+b^2}(a \cos bu + b \sin bu) + C$

55. $\int \ln u\, du = u \ln u - u + C$

56. $\int \dfrac{1}{u \ln u}\, du = \ln|\ln u| + C$

57. $\int u^n \ln u\, du = \dfrac{u^{n+1}}{(n+1)^2}[(n+1)\ln u - 1] + C$

58. $\int u^m \ln^n u \, du = \dfrac{u^{m+1} \ln^n u}{m+1} - \dfrac{n}{m+1} \int u^m \ln^{n-1} u \, du, \quad m \neq -1$

59. $\int \ln(u^2 + a^2) \, du = u \ln(u^2 + a^2) - 2u + 2a \tan^{-1} \dfrac{u}{a} + C$

60. $\int \ln |u^2 - a^2| \, du = u \ln |u^2 - a^2| - 2u + a \ln \left| \dfrac{u+a}{u-a} \right| + C$

61. $\int \sinh u \, du = \cosh u + C$ **62.** $\int \cosh u \, du = \sinh u + C$

63. $\int \tanh u \, du = \ln \cosh u + C$ **64.** $\int \coth u \, du = \ln |\sinh u| + C$

65. $\int \operatorname{sech}^2 u \, du = \tanh u + C$ **66.** $\int \operatorname{csch}^2 u \, du = -\coth u + C$

67. $\int \operatorname{sech} u \tanh u \, du = -\operatorname{sech} u + C$ **68.** $\int \operatorname{csch} u \coth u \, du = -\operatorname{csch} u + C$

69. $\int \sqrt{a^2 + u^2} \, du = \dfrac{u}{2} \sqrt{a^2 + u^2} + \dfrac{a^2}{2} \ln \left| u + \sqrt{a^2 + u^2} \right| + C$

70. $\int u^2 \sqrt{a^2 + u^2} \, du = \dfrac{u}{8} (a^2 + 2u^2) \sqrt{a^2 + u^2} - \dfrac{a^4}{8} \ln \left| u + \sqrt{a^2 + u^2} \right| + C$

71. $\int \dfrac{\sqrt{a^2 + u^2}}{u} \, du = \sqrt{a^2 + u^2} - a \ln \left| \dfrac{a + \sqrt{a^2 + u^2}}{u} \right| + C$

72. $\int \dfrac{\sqrt{a^2 + u^2}}{u^2} \, du = -\dfrac{\sqrt{a^2 + u^2}}{u} + \ln \left| u + \sqrt{a^2 + u^2} \right| + C$

73. $\int \dfrac{du}{\sqrt{a^2 + u^2}} = \ln \left| u + \sqrt{a^2 + u^2} \right| + C$

74. $\int \dfrac{u^2 \, du}{\sqrt{a^2 + u^2}} = \dfrac{u}{2} \sqrt{a^2 + u^2} - \dfrac{a^2}{2} \ln \left| u + \sqrt{a^2 + u^2} \right| + C$

75. $\int \dfrac{du}{u \sqrt{a^2 + u^2}} = -\dfrac{1}{a} \ln \left| \dfrac{\sqrt{a^2 + u^2} + a}{u} \right| + C$

76. $\int \dfrac{du}{u^2 \sqrt{a^2 + u^2}} = -\dfrac{\sqrt{a^2 + u^2}}{a^2 u} + C$

77. $\int \sqrt{u^2 - a^2} \, du = \dfrac{u}{2} \sqrt{u^2 - a^2} - \dfrac{a^2}{2} \ln \left| u + \sqrt{u^2 - a^2} \right| + C$

78. $\int u^2 \sqrt{u^2 - a^2} \, du = \dfrac{u}{8} (2u^2 - a^2) \sqrt{u^2 - a^2} - \dfrac{a^4}{8} \ln \left| u + \sqrt{u^2 - a^2} \right| + C$

79. $\int \dfrac{\sqrt{u^2 - a^2}}{u} \, du = \sqrt{u^2 - a^2} - a \cos^{-1} \dfrac{a}{u} + C$

80. $\int \dfrac{\sqrt{u^2 - a^2}}{u^2} \, du = -\dfrac{\sqrt{u^2 - a^2}}{u} + \ln \left| u + \sqrt{u^2 - a^2} \right| + C$

81. $\displaystyle\int \frac{du}{\sqrt{u^2 - a^2}} = \ln\left|u + \sqrt{u^2 - a^2}\right| + C$

82. $\displaystyle\int \frac{u^2\,du}{\sqrt{u^2 - a^2}} = \frac{u}{2}\sqrt{u^2 - a^2} + \frac{a^2}{2}\ln\left|u + \sqrt{u^2 - a^2}\right| + C$

83. $\displaystyle\int \frac{du}{u^2\sqrt{u^2 - a^2}} = \frac{\sqrt{u^2 - a^2}}{a^2 u} + C$

84. $\displaystyle\int \frac{du}{(u^2 - a^2)^{3/2}} = -\frac{u}{a^2\sqrt{u^2 - a^2}} + C$

85. $\displaystyle\int \sqrt{a^2 - u^2}\,du = \frac{u}{2}\sqrt{a^2 - u^2} + \frac{a^2}{2}\sin^{-1}\frac{u}{a} + C$

86. $\displaystyle\int u^2\sqrt{a^2 - u^2}\,du = \frac{u}{8}(2u^2 - a^2)\sqrt{a^2 - u^2} + \frac{a^4}{8}\sin^{-1}\frac{u}{a} + C$

Some Basic *Mathematica* Syntax and Commands

■ *Mathematica* conventions

■ Most implementations of *Mathematica* consist of two parts, the *front end* and the *kernel*. The lines you type into the Front End of *Mathematica* are called inputs and are shown in **boldface type** (unless you specify another convention). The output from *Mathematica* is generated by the kernel and is shown in plainface type. Generally, the output is displayed unless you specifically suppress it by using a semicolon. The construct (*** comment ***) is ignored by the kernel and can be used to annotate inputs in the front end.

In[1]:= **2**

Out[1]= 2

In[2]:= **3;** (*** The corresponding output of 3 is suppressed, and Out[2] is not shown. ***)

In[3]:= **4; 5** (*** The 4 is suppressed and the 5 is shown. ***)

Out[3]= 5

■ An equal sign (=) is used for assignment.

In[4]:= **a=5**

Out[4]= 5

In[5]:= **b=3;**

In[6]:= **b** (*** Just to be sure that b was correctly input. ***)

Out[6]= 3

In[7]:= **a+b**

Out[7]= 8

In[8]:= **ab** (*** This is the variable ab, not the product of a and b. ***)

Out[8]= ab

In[9]:= **a b** (*** When a and b are separated by a space we have the product of a and b. ***)

Out[9]= 15

■ Grouping is done with parentheses only, not with square backets [ ] or braces { }.

In[10]:= **a/(a+b)**

Out[10]= $\frac{5}{8}$

■ Arguments of functions are enclosed in square brackets.

In[11]:= **Sqrt[9]**

Out[11]= 3

■ Braces are used for ordered sets - called *lists* in *Mathematica*. The individual elements of a list are obtained using double square brackets [[ ]].

In[12]:= **m = {2, 7, -8}**

Out[12]= {2, 7, −8}

In[13]:= **m[[2]]**

Out[13]= 7

■ Built-in constants

■ π, E, I, Infinity

In[14]:= π (* **This is input using option-p. Equivalently, Pi can be typed.** *)

Out[14]= π

In[15]:= **N[π]** (* **N is a function than converts a number to decimal form.** *)

Out[15]= 3.14159

In[16]:= **E//N** (* **Using //, a function can be typed following its argument.** *)

Out[16]= 2.71828

In[17]:= **I^2** (* **I is the square root of** −1. *)

Out[17]= −1

In[18]:= **Infinity**

Out[18]= ∞

■ Built-in mathematical functions

■ The following functions are built into *Mathematica* with the syntax shown.

absolute value: **Abs[x]**
square root: **Sqrt[x]**
trigonometric: **Sin[x], Cos[x], Tan[x], Cot[x], Sec[x], Csc[x]**
inverse trigonometric: **ArcSin[x], ArcCos[x], ArcTan[x], ArcCot[x], ArcSec[x], ArcCsc[x]**
exponential: **Exp[x]** or **E^x**
natural logarithm: **Log[x]**
hyperbolic: **Sinh[x], Cosh[x], Tanh[x], Coth[x], Sech[x], Csch[x]**

In[19]:= **Sqrt[−1]**

Out[19]= i

In[20]:= **Sqrt[90]**

Out[20]= $3\sqrt{10}$

In[21]:= **Sqrt[90]//N**

Out[21]= 9.48638

In[22]:= **N[Sin[Pi/3]]**

Out[22]= 0.866025

In[23]:= **Log[5]**

Out[23]= Log[5]

In[24]:= **Log[5]//N**

Out[24]= 1.60944

In[25]:= **Log[Exp[6]]**

Out[25]= 6

In[26]:= **Sinh[Log[3]]**

Out[26]= $\dfrac{4}{3}$

■ Derivatives

■ The syntax for finding the derivative of a function is discussed in Section 1.1 of this manual. Below we show how *Mathematica* can be used to verify that $2x/(3 - x^2)$ is a solution of the differential equation $xy' - y = xy^2$.

In[27]:= **Clear[y]**
lhs = x y′[x] - y[x]
rhs = x y[x]^2

Out[28]= $-y[x] + x\, y'[x]$

Out[29]= $x\, y[x]^2$

In[30]:= **y[x_]:=2x/(3 − x^2)**

y[x]

Out[31]= $\dfrac{2x}{3 - x^2}$

In[32]:= **lhs − rhs//Simplify**

Out[32]= 0

■ Rules

■ We saw above under ***Mathematica* conventions** how to assign a specific value to a variable. For example, if we want the variable x to have the value 2 we input **x=2**. Once this is done, until x is either reassigned or cleared, it will have the value 2. Suppose, on the other hand, we have an expression containing x, such as $x^2 + 2x - 5$, and we want the value of this expression when $x = 2$, but we do not want to permanently assign the value 2 to x. One way to do this is to define the expression as a function **f** and evaluate **f[2]**. However, if we are only going to do this for a single value of x, a faster way is to use the following syntax.

In[33]:= **x^2 + 3x - 1 /. x ->2**

Out[33]= 9

The arrow from x to 2 is obtained by typing a *minus sign* followed by the *greater than* symbol. The expression **x ->2** is called a rule and **/.** is a replacement operator. The complete input expression **x^2 + 3x - 1 /. x ->2** says to replace x with 2 in $x^2 + 3x - 1$, but do not permanently assign the value 2 to x.

In[34]:= **x+3 (* x has not been permanently assigned the value 2. *)**

Out[34]= 3+x

■ Graphing with color and dashing

■ In Section 2.2 of this manual we saw that the **Plot** command can be supplemented with options like **PlotRange** to modify the appearance of a graph. *Mathematica* also has options for colorizing graphs and drawing graphs with dashed lines or thicker lines. The **RGBColor** option is used to obtain graphs of various colors. Red is **[1,0,0]**, green is **[0,1,0]**, and blue is **[0,0,1]**. Other colors are obtained by using numbers between 0 and 1 as the arguments of **RGBColor**. On a black and white printer the colors will appear as shades of gray as seen below.

In[35]:= **Plot[{ x^2, x, Sin[x], 2.5 - x^2}, {x,-3,3}, PlotStyle – > { RGBColor[1,0,0], RGBColor[0,1,0], RGBColor[0,0,1], RGBColor[1,0.5,0.8]}]**

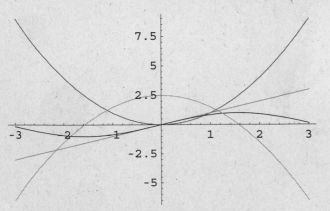

Out[35]= -Graphics-

■ The example below shows how to obtain a dashed graph, and thicker graphs in various shades of gray. The arguments for **Dashing** specify, alternately, the length of the dash and the length of the open space. The argument for **Thickness** is a number from 0 to 1 where 1 represents the width of the entire plot. The argument for **GrayLevel** is also a number from 0 to 1 where 0 is black and 1 is white.

In[36]:= **Plot[{ x^2, x, Sin[x], 2.5 - x^2},{ x,-3,3}, PlotStyle- > {{ Dashing[{0.05,0.05}],**

{ Dashing[{0.01,0.06}], { Thickness[0.015], GrayLevel[0.4]},

{ Thickness[0.025], GrayLevel[0.8]}]

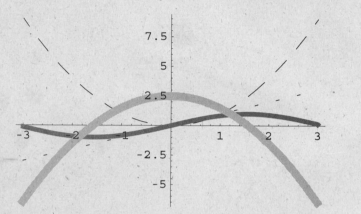

Out[36]= -Graphics-